Palgrave Studies in Water Governance: Policy and Practice

Series Editors
Christian Bréthaut
Institute for Environmental Sciences
University of Geneva
Geneva, Switzerland

Thomas Bolognesi
GEDT
University of Geneva
Geneva, Switzerland

Looking at the issues of water governance through the perspective of the social sciences, books in the Palgrave Series in Water Governance take a global perspective on one of the key challenges facing society today: the sustainable development of water resources and services for all. In stepping away from the traditional focus on engineering and geophysics, the series takes a more holistic approach to both consolidate and generate knowledge that can be applied to different geographic areas by academics, researchers, policy-makers, NGOs and the private sector. This series emphasises the link between science and policy through considering water as a socio-ecological system, water and the territoriality of action, and water in the context of conflicts.

More information about this series at
http://www.palgrave.com/gp/series/15054

Manuel Fischer • Karin Ingold
Editors

Networks in Water Governance

palgrave
macmillan

Editors
Manuel Fischer
Department of Environmental
Social Sciences, Eawag
Dübendorf, Switzerland

Karin Ingold
Institute of Political Sciences
University of Bern
Bern, Switzerland

Palgrave Studies in Water Governance: Policy and Practice
ISBN 978-3-030-46768-5   ISBN 978-3-030-46769-2   (eBook)
https://doi.org/10.1007/978-3-030-46769-2

This Palgrave Macmillan imprint is published by the registered company Springer Nature Switzerland AG.
The registered company address is: Gewerbestrasse 11, 6330 Cham, Switzerland

Acknowledgement

The book editors would like to thank Aline Hänggli for her tremendous support with coordination and layout.

Contents

Notes on Contributors

Mario Angst is an early career researcher from Zurich, Switzerland. His research covers environmental governance and sustainability policy, often from a network perspective. Among other things, this has meant investigating organizational networks in Swiss water governance, network development in regional nature parks, and also policy instrument choice in renewable energy transitions. He is working on advancing social-ecological network models for ecosystem governance at the Swiss Federal Institute for Forest, Snow and Landscape Research WSL.

Emily V. Bell is Assistant Professor of Public Administration and Policy at the University of Georgia, with a PhD from the University of Arizona. Her research uses inferential and descriptive social network analysis to examine policy coordination and collaborative processes in water governance. Bell's recent work as a postdoctoral associate at the Duke University Nicholas School of the Environment has examined these factors in the context of climate hazard mitigation, focusing on extreme drought and precipitation, as well as sea level rise.

Örjan Bodin is an associate professor at the Stockholm Resilience Centre at Stockholm University. He is studying different challenges and opportunities in governing natural resources, and much of his research bridges the natural and social sciences.

Hans Th. A. Bressers is Professor of Policy Studies and Environmental Policy at the University of Twente in the Netherlands and founder of the CSTM, currently the Department of Governance and Technology for Sustainability. Next to various other national advisory roles, until 2018, he also was a member of the Dutch Advisory Committee on Water that was chaired by the Prince of Orange until April 2013 when he became king.

Stephanie Bultema is a PhD student in the School of Public Affairs and a researcher in the Center on Network Science at the University of Colorado Denver. She uses network science to advance knowledge of inter-organizational collaboration in theory and practice. Her research focuses on the collaborative governance of community health system transformation.

Fariba Ebrahimiazarkharan holds a PhD in Natural Resources Management and Water Governance from the University of Tehran, Iran, at the College of Agriculture and Natural Resources. Her main research expertise focuses on novel methods to water management and governance in natural resources and rivers.

Manuel Fischer is a research group leader at the Department of Environmental Social Sciences at the Swiss Federal Institute for Water Science and Technology (Eawag) and a lecturer at the Institute of Political Science at the University of Bern, Switzerland. His research deals with water and environmental governance and policy-making.

Mehdi Ghorbani is an associate professor at the University of Tehran, Iran. His main research interests, teaching, and outreach activities focus on the governance of natural resources, social network analysis, social-ecological systems modeling, building resilience, and local communities empowerment. He spent his sabbatical leave at the professorship of environmental policy and economics at ETH Zurich in Switzerland between 2010 and 2011.

Adam Douglas Henry is a distinguished scholar and Professor of Environmental Politics at the University of Arizona. His research applies network analysis and computational modeling to the study of the policy process, particularly surrounding issues of environment and sustainability. Henry's work has been applied to the study of renewable energy, environmental risk, urban water sustainability, and invasive species management. Henry received his PhD from the University of California, Davis, USA, and was a Giorgio Ruffolo Fellow in Sustainability Science at the Kennedy School of Government, Harvard University.

Laura Mae Jacqueline Herzog is a postdoctoral fellow at the Chair of Resource Management at the Institute of Environmental Systems Research at the University of Osnabrück. In her research, she investigates the interactions of humans with water and land resources and the influences of these interactions on the respective ecological system. She is interested in the potential conflicts of interest emerging from such human-ecosystem interactions and how their influences on nature can be shaped sustainably.

James Hollway is Associate Professor of International Relations/Political Science at the Graduate Institute of International and Development Studies, Geneva. There he co-directs the Professional Skills Training program, and he is affiliated with the Centre for International Environmental Studies, the Centre for Trade and Economic Integration, and the Global Governance Centre. He is also a fisheries cluster leader in the Taskforce on Oceans Governance of the Earth Systems Governance research alliance. His research focuses on international institutions, especially those governing fisheries, water, or trade, and develops network theories and methods for studying their interdependencies.

Karin Ingold is a professor at the Institute of Political Science and the Oeschger Center for Climate Change Research at the University of Bern. She is also affiliated to the Environmental Social Science Department at the Swiss Federal Institute for Water Science and Technology (Eawag). She is interested in policy design and politics of climate change, water, and energy.

Elizabeth A. Koebele is Assistant Professor of Political Science at the University of Nevada, Reno. She holds a PhD in Environmental Studies from the University of Colorado Boulder. Her research focuses on the role of collaboration in environmental governance processes, with an emphasis on water and natural hazards management.

Mark Lubell is a professor at the Department of Environmental Science and Policy and the director of the Center for Environmental Policy and Behavior at the University of California, Davis, USA. He is interested in cooperation problems and decision-making in environmental, agricultural, and public policy. Field research topics include water management, sustainable agriculture, adaptive decision-making, climate change policy, local government policy, transportation behavior, plant disease management, invasive species, and policy/social network analysis.

Arash Malekian is an associate professor at the University of Tehran, Iran. He dedicates his research and teaching to hydrology and water resources management, climate change, mitigation, policy, and decision-making. He has served as the advisor of the international office at the College of Agriculture and Natural Resources, and as the head of department and international advisor of Science and Technology Park of the University of Tehran.

María Mancilla García is a researcher at the Stockholm Resilience Centre in Sweden. She works on relational approaches to the study of social-ecological systems. Her work ranges from applying process-relational philosophy to the analysis of social-ecological systems as complex adaptive systems to conducting empirical studies on water governance and fisheries governance in Latin America and Africa.

Tomás Olivier is an assistant professor in the School of Public Administration, Florida Atlantic University. His research focuses on the role and design of institutions for addressing collective action problems surrounding water governance.

Matthew Robbins is an environmental social scientist and holds a PhD in Ecology from the University of California, Davis. He has studied collaborative environmental policy arenas, both international and domestic, through a number of qualitative and quantitative methods. He also has applied experience working as a Sea Grant Fellow at the Delta Stewardship Council. Robbins is a post-doctoral scholar in the UC Davis Computational Communication Research Lab, studying the science of science communication using natural language processing and network analysis.

Edella Schlager is Professor of Public Policy and Director of the School of Government and Public Policy at the University of Arizona. Her research focuses on river basin governance in the USA, with a particular emphasis on polycentric institutions. She also examines the vulnerability and robustness of water institutions to climate change impacts. She is the co-author of several books on water governance, and her articles have appeared in the *American Journal of Political Science*, *Policy Studies Journal*, *Public Administration Review*, and *Land Economics*.

Tyler A. Scott is an assistant professor in the Department of Environmental Science and Policy and the co-director of the Center for Environmental Policy and Behavior at the University of California, Davis.

Christopher M. Weible is a professor in the School of Public Affairs at the University of Colorado Denver. He co-directs the Workshop on Policy Process Research and studies conflict and concord in policy processes.

List of Figures

List of Tables

1

Introduction

Manuel Fischer and Karin Ingold

There is hardly any other natural resource as inherently complex as water. Water provides a large diversity of ecosystem services (irrigation, cooling, drinking, support for biodiversity, hydropower, etc.) that are often in conflict with each other. Furthermore, water is an important resource for other sectors such as health, energy, or agriculture. With increasingly important consequences of climate change and biodiversity loss, not only the protection of water resources and water-related ecosystems becomes increasingly important, but also the protection *from* water gains importance as well, for example, through flood protection measures. Such measures, in turn, can have consequences again for drinking water provision, agricultural land use, or the state of ecosystems (Jaramillo et al. 2019). Thus, water governance is complex, and network concepts and measures

M. Fischer (✉) • K. Ingold
Federal Institute of Aquatic Science and Technology, Eawag,
Dübendorf, Switzerland

University of Bern, Bern, Switzerland
e-mail: manuel.fischer@eawag.ch; karin.ingold@ipw.unibe.ch

© The Author(s) 2020
M. Fischer, K. Ingold (eds.), *Networks in Water Governance*, Palgrave Studies in Water
Governance: Policy and Practice, https://doi.org/10.1007/978-3-030-46769-2_1

allow us to grasp some of that complexity as well as ways of addressing it (Angst 2018).

Due to its substantive importance for life, and due to the interest in studying this sector given the different challenges that arise therein, water has always been a key policy sector for studies of the policy process, public administration, and environmental governance. Research on governance and management of water resources has thus traditionally contributed to new developments in the broader fields of public policy, public administration, and governance studies, and beyond (Weible and Sabatier 2005; Feiock and Scholz 2009; Berardo and Scholz 2010; Pahl-Wostl et al. 2010; Lubell 2013; Sabatier and Weible 2014; Bodin 2017; Berardo and Lubell 2019).

Not surprisingly, a significant proportion of these approaches have relied, explicitly or implicitly, on concepts and measures related to networks. The governance of complex resources such as water calls for a variety of relevant actors to interact, which ideally allows one to take into account the different usage- and protection-related interests toward water, coordinate across politico-administrative boundaries, and include knowledge from different sectors, including public, private, as well as scientific actors (Maag and Fischer 2018). These interactions across diverse categories of actors can be fruitfully analyzed as networks, and with related concepts and measures. Furthermore, because formal political institutions have a hard time addressing issues that span political or sectoral borders, as is the case with water, (informal) networks of collaboration and information exchange among actors are even more important (Galaz 2007; Ingold et al. 2016; Lubell and Edelenbos 2013). Networks are thus not only a relevant conceptual and methodological approach to analyzing water governance, they are also an empirical reality, and often an aspirational goal of stakeholders that want to address pressing water-related issues in a holistic fashion.

Why This Book?

Currently, water governance and management is facing severe challenges due to climate change, biodiversity loss, increasing urbanization, and/or population growth. We are confident that by studying water with the help of network concepts and measures, we can advance our knowledge in these arenas and provide a contribution to successfully tackle these current and future challenges.

The book addresses the issue of water and its management and regulation from a holistic perspective. This means that it covers both the complexities in terms of the different issues, aspects, policy problems, policy sectors, and substantive questions related to water governance, and the complexities that the involved set of actors (state officials, public administrations, scientists, non-governmental organizations (NGOs), private firms, lobbying groups, and others) are facing when dealing with the overlapping issue of water. Whereas different chapters deal with different water-related issues, from more specialized issues such as fisheries or water quality to more general issues such as watershed management or subsystem governance, all chapters deploy the same particular lens of analysis. They all rely on a network perspective and apply specific concepts and measures of Social Network Analysis (SNA) to grasp the complex interactions among the large diversity of actors, institutions, and issues related to water governance. Bodin and Prell (2011) have emphasized the importance of analyzing environmental and natural resource questions based on concepts and methods of network analysis. Whereas all chapters deal with water and involve a network approach, their theoretical focus varies, although network concepts per se could themselves be regarded as the theoretical background in many chapters. Similar books that rely on one single theoretical framework are Feiock and Scholz (2009) with the Institutional Collective Action Framework, or Weible et al. (2016), featuring country-focused chapters dealing with hydraulic fracturing and applying the Advocacy Coalition Framework.

This book is relevant to different types of academic audiences, including students, researchers, and teachers interested in natural resource governance and environmental policy-making. These issues are increasingly important in many university departments and related curricula, including interdisciplinary studies. The book will thus be of interest for students interested in water-policy and environmental governance; it provides input appropriate for introductory classes as well as resources and examples for more specialized seminars. Presumably, Chap. 2 where conceptual foundations related to water governance and networks are presented can serve as an important introductory text for teaching. The nine case study chapters represent first-hand examples of water-related policy and governance issues, but also of high-level academic research on the issues, written by some of the leading water governance and network scholars in the world. Finally, our concluding considerations on comparability, generalizability, caveats, and related open research questions can provide inspiration to the current and future generations of water governance scholars and SNA practitioners.

Besides academic contributions, this book also aims to provide practice-relevant recommendations. To gather network data, we are dependent on information from practice: only through the access to official documents and expert judgments are we able to draw networks between actors, institutions, and issues. Designing a network of actors in charge of managing the resource water also means highlighting lacunae where important links are missing. There are more and less efficient and effective ways of collaborating, interacting with each other, or exchanging information. This book sheds light on how different actors in different countries, regions, from different sectors, and related to different water issues, interact. We hope that these examples—and our subsequent discussion in the conclusion—provide practitioners with new ideas on what approaches can work in what contexts. Thus, the practical implications discussed in the chapters and at the end of the book will hopefully be relevant to scholars that aim to have an impact beyond academia, and, more importantly, to many water-related practitioners in public administrations, NGOs, international organizations, and so on.

Summary of Case Study Chapters

This book focuses on different aspects of water governance through the lenses of network concepts and measures. Besides this introduction, a conceptual chapter, and the conclusion, this book contains nine case study chapters that each emphasize different aspects of water governance and rely on different network concepts and measures. The first three case study chapters focus on network fragmentation and clustering within networks of water governance. The subsequent three case study chapters present ways to overcome this fragmentation and reduce clustering. Finally, the last three case study chapters focus on centrality, and thus on specific actors that drive water governance (see Table 1.1).

This introduction first lays out the rationale of the book. It then provides summaries of all chapters, before offering an overview of the substantive water issues covered by the chapters, the geographical areas concerned, the network ties analyzed, and the network concepts and measures that each chapter relies on. It finally discusses the different research questions addressed by each chapter and presents the structure of the overall book.

Chapter 2 lays the conceptual foundations for the book and discusses water governance, network concepts and methods, and their interrelationship. It does so by referring to the case study chapters and the elements studied therein. After systematically presenting the complexities of water governance and the ways water issues are typically governed and managed, the chapter discusses the contributions of network concepts to understanding these complexities, as well as ways to overcoming problems that arise from such complexities. The chapter then systematically presents types of nodes and ties of networks, before providing a short explanation of the most basic network analytical concepts on micro-, meso- and macro-levels of networks. It finally covers certain methods for statistical network analysis and discusses the most common challenges and shortcomings related to network analytical approaches.

In the first empirical case study chapter, Mateo Robbins and Mark Lubell provide a longitudinal analysis of the Honduran Spiny Lobster Fishery Governance Network. As conservation initiatives have grown in

Table 1.1 Chapter overview

Chapter authors	Water issue	Case study region	SNA ties	SNA concepts and measures
2.1 Fragmentation in water politics				
2.1.1 Robbins & Lubell	Conservation/Spiny lobster fishery	Honduras, Caribbean	Communication, familiarity	Network segregation and social capital
2.1.2 Hollway	International water treaties	Global	Conflict, cooperation	Activity and network change over time
2.1.3 Angst & Fischer	Swiss Water Policy	Switzerland	Actor—issue involvement (bipartite)	Modularity; two-mode centrality
2.2 Fostering collaboration in complex systems				
2.2.1 Mancilla & Bodin	Water basin management	Brazil	Coordination, shared beliefs	Homophily and heterophily
2.2.2 Koebele et al.	Water basin management	USA, Lake Tahoe	Collaboration, shared beliefs	Polarization, centrality
2.2.3 Herzog & Ingold	Water quality, micro-pollutants	Rhine catchment: Basel, Moselle, Ruhr	Collaboration	Cohesion, centrality, clustering
2.3 Key actors in networks				
2.3.1 Ebrahimiazarkharan et al.	Water quality and quantity	Iran, Thalaghan watershed	Cooperation, trust	Cohesion, centrality, clustering
2.3.2 Olivier et al.	Watersheds management	USA, New York watershed	Actors—protocol (bipartite), joint protocol	Network rules and social capital
2.3.3 Bell & Henry	Water sustainability	USA, Arizona	Event co-participation	Centralities

scope and scale, the challenges of scaling up collaborative conservation efforts among a broader and more diverse base of participants, while at the same time managing their multiple and often competing interests, have become more acute. These initiatives seek to catalyze governance networks and to broaden participation and communication among stakeholders. The chapter analyzes the evolution of the governance network in the Caribbean spiny lobster fishery of Honduras over the course of the Spiny Lobster Initiative (SLI). It first presents a set of five mechanisms behind network fragmentation and segregation, and formulates related hypotheses of why a given actor would connect with certain other actors. Based on community detection, the chapter analyses to what degree network communities overlap with attributes that would cause network segregation. The authors then apply stochastic actor-oriented modeling (SAOM) to their longitudinal network data.

The second empirical case study chapter by James Hollway focuses on international water cooperation. The geography of international water basins strongly determines countries' potential water agreement partners. The chapter analyzes whether cooperative activity begins with the most vulnerable downstream states first, and then spreads through bilateral or multilateral agreements to cover the entire basin. Further, it asks whether existing partners choose to develop further agreements with each other only once agreements with all other basin parties are established. Finally, it examines whether the institutional design of existing agreements matters for the diffusion of institutional water agreements. Drawing on literature on (international environmental) institutional negotiation and design, the chapter theorizes factors affecting the timing of institutional agreement, including certain features of the current state of their local network. The analysis relies on the International Environmental Agreements dataset, complemented by some additional coding, giving over 700 bilateral and multilateral water agreements, and affording the chapter a global scope. To estimate states' activity rates in concluding international water agreements, the analysis employs Dynamic Network Actor Models (DyNAMs).

The case study chapter by Mario Angst and Manuel Fischer deals with subsystem identification and crucial actors in Swiss water politics. The water policy system is represented as a two-mode network of actors and issues, but this system is still subdivided into different, albeit connected subsystems that deal with different water-related issues. Actors that connect issues and other actors within or across policy sectors are important because they can act as brokers between issues. The empirical study includes all possible issues that are relevant in water politics and related fields, such as hydropower, wastewater, energy, agriculture, and many more. The dataset, based on a survey among over 300 actors, covers Swiss water politics on a national as well as a regional level. In terms of Social Network Analysis, this chapter emphasizes the concept of two-mode centrality, that is, the centrality of actors in terms of the issues they deal with. The study first applies a modularity measure to subdivide the entire network into five subsystems. It then relies on two-mode centrality measures adopted from ecological studies in order to identify within-subsystem connectors as well as between-subsystem connectors, and discusses illustrative examples of these actors.

The case study chapter by María Mancilla García and Örjan Bodin focuses on network patterns among participants of a water basin council in Brazil. Participatory forums for water governance have been created all over the world with the hope that facilitating communication among actors who did not previously communicate would enhance sustainable governance of water resources. Such forums can thus potentially help overcome fragmentation in water politics. In Latin America, these forums have been implemented across countries, presenting variable sizes and time spans of existence. The empirical analysis deals with the council of the Paraiba do Sul federal river basin in Brazil that crosses several federal states. The chapter explores whether participants of the forum tend to communicate with other participants that are similar to them in terms of policy sectors, state identification, or beliefs on water governance, or whether actors tend to approach those they see as most powerful. These hypotheses are then assessed based on Quadratic Assignment Procedures (QAP) and Exponential Random Graph Models (ERGMs), two tools for the statistical analysis of network data. The discussion of results is combined with qualitative insights from interviews.

Elizabeth Koebele, Stephanie Bultema, and Christopher Weible analyze polarizing and converging discourses around policy change related to the management of Lake Tahoe in California. Political interactions among policy actors are manifested in many ways, including through their discourses. As actors seek to influence, negotiate, and react to policy changes, their discourses evolve to reflect greater levels of polarization or convergence. Guided by the theoretical framework of the Advocacy Coalition Framework (ACF), the chapter analyses the discourse of policy actors in the Lake Tahoe Basin in California, which contains one of the most pristine alpine lakes in the world. The time interval covered in the analysis is before and after a major policy change: the adoption of the 2012 Regional Plan. Relying on Discourse Network Analysis, the authors analyze almost 100 newspaper articles across three publication outlets that discuss the design, adoption, and aftermath of the Plan from 2005–2014. The authors then compare three networks, one reflecting actors' beliefs, one reflecting their policy positions, and one pertaining to their coordinated or unintentional interactions. The analysis finally compares the structure of the different networks, and assesses which types of actors are most central according to the different types of networks.

The chapter by Laura Herzog and Karin Ingold deals with efficient transboundary water management and the role of water body organizations. (Micro-)pollution in river surface waters does not stop at borders, and upstream-downstream dynamics inherent to water necessitate a transboundary river water management. Enhancing communication and fostering trust among actors from different sectors and distinctive jurisdictions can be important to activating collective action. The study relies on the Social-Ecological Systems (SES) framework and deals with water quality management and drinking water supply in the Rhine catchment area, and compares coordination patterns between Swiss, German, and Luxembourgian actors involved in water quality regulation. More specifically, the authors analyze networks of actor collaboration in three sub-catchments of the Rhine River and compare these networks in terms of macro-, meso-, and micro-level network statistics.

Fariba Ebrahimiazarkharan, Mehdi Gorbani, Arash Malekian, and Hans Bressers identify challenges and opportunities in the water governance network of an Iranian watershed and focus on the role of key actors therein. Various actors at different levels are engaged with the water governance of the watershed. The analysis of social ties of cooperation in the local community network, and the pinpointing of related opportunities and threats helps to advance adaptive governance of water resources in the region. Based on fieldwork, interviews, and visits by governmental organizations, the authors identify 13 villages out of 82 with very important problems related to water resource management. Three hundred and ninety local beneficiaries or water users were identified based on snowball sampling, and interviews and surveys allowed to gather cooperation ties among them. This network is then analyzed at the levels of the entire network (macro-structures), of subgroups (meso-structures), and of the position of individual actors in the network (micro-structures).

The chapter by Tomás Olivier, Tyler Scott, and Edella Schlager deals with assessing the structure of rule networks for resolving collective action problems. Much of the network scholarship is highly social in focus, examining relationships such as information sharing, trust, and regular communication between actors using a social capital framework. Rules and governance protocols are well recognized as critical drivers of collective action situations, but have largely been left out of social network-oriented analyses. This chapter bridges this gap by analyzing the structural arrangement of institutional rules specifying inter-actor relationships and behaviors using a social capital framework (e.g., bridging and bonding structures). The authors analyze the network of four thousand rules constituting the New York City Watersheds governing arrangement that provide for sixty public goods, eleven shared decision making venues, and dozens of monitoring and enforcement mechanisms. Overall, the chapter explores promising approaches that allow for the combined consideration of institutions and social interactions.

Finally, Emily Bell and Adam Henry compare centrality in policy networks with centrality in media narratives in local water policy. Social network analysis provides useful tools for the simplification of complex policy systems, such as urban water governance, as well as the measurement of phenomena of theoretical importance in policy studies. The chapter considers two of the most widely used measures of actor position in networks analysis—degree centrality and betweenness centrality—and analyzes the degree to which these measures capture the theoretical concepts of policy leadership, entrepreneurship, and brokerage. The empirical context is local water governance in a single policy subsystem in Tucson, Arizona, a semi-arid desert municipality, where many policy processes are characterized by fragmentation and political conflict. Systematic analysis of media reports and grey literature is used to identify policy events surrounding water sustainability across multiple domains such as flood management, water quality, and water supply, and actors participating in these events. This results in a two-mode network of actors and policy events, from which a unipartite network of actor-actor connections is created. The authors then correlate centrality in this network with narrative accounts of the importance of these actors, as well as the narrative classification of these actors as policy brokers, policy entrepreneurs, and leaders.

The last chapter presents the conclusions of this book. It summarizes the main findings of each chapter and puts them in relation to the overarching concepts discussed in the introduction and the chapter on conceptual foundations (Chap. 2). It then presents a research agenda mapping out relevant questions for the future, and discusses some practical insights for water governance and policy practitioners.

Comparative Overview of Case Study Chapters

Each of the case study chapters contains three key elements. First, it deals with a specific water-related issue on the political agenda. Second, each chapter focuses on a case study area, that is, a region, a local governance situation, or a country. Third, each chapter studies the water-related issue in this case study area through a specific type of network and at least one network concept, operationalized through Social Network Analysis (SNA). These elements are combined in order to answer a specific research question. Table 1.1 presents an overview of the case study chapters and their key elements.

As can be grasped from the overview in Table 1.1, the case study chapters together cover a wide range of water issues, case study regions, and types of network ties and related network concepts and methods. In terms of water issues, some chapters deal with specific issues such as lobster fishing and related ecosystem protection or with micro-pollutants in river streams. Other chapters address water governance in general and the interaction of different issues, such as within a watershed, a lake area, a municipality, an entire country, or on the level of international cooperation among states. They cover cases as diverse as the USA, Switzerland, Brazil, Germany, Honduras, and Luxembourg.

Network ties analyzed in the chapters include collaboration, coordination, or communication among actors, influence attribution among actors, shared believes among actors, joint responsibility for an issue due to formal laws, event co-participation, or conflict. These networks are analyzed with concepts and measures such as network clustering, different types of network centralities, network density, centralization or transitivity, core-periphery structures of networks, as well as statistical models that link network (sub)structures to a diversity of exogenous and endogenous explanatory factors.

All of these key elements in the case study chapters are combined in order to answer a range of relevant research questions. In summary, these research questions can be subdivided into two general queries. On the one hand, authors ask what actors are in particularly important, central positions within networks, and how such central positions are distributed within networks or parts thereof, such as subsystems or coalitions. These questions are important as they deal with how to identify actors that are particularly powerful in networks, or that are in a position to act as brokers between different parts of the networks. On the other hand, authors ask why actors in water governance networks interact with some actors, but not with others. Different explanatory factors such as geographical location or sectoral identities of actors, trust and communication options among actors, or actors' decisions made based on existing network structures are taken into account in finding the answer. These questions, or similar ones, are related to the issue of fragmentation of water governance networks, and the discussion of how to overcome this fragmentation.

The aforementioned questions are related to the networks directly; however, the authors also engage broader questions, which nicely illustrate the ways in which network analytic concepts and measures can potentially contribute to larger discussions around participation in governance and policy-making or the quality of policy outputs. Examples are "How do networks influence the provision of public goods?", "How can adaptive governance be achieved by fostering networks?", "How can networks improve social and environmental outcomes in natural resource governance?", "How do actors and their networks contribute to policy innovation?", "How can more collaborative and equitable governance arrangements be achieved?" These questions all relate to larger political concepts such as efficiency, quality, or legitimacy of governance and collective decision-making, and do a good job of illustrating the normative goals that are frequently pursued by both practitioners and researchers in water governance.

Even though all chapters rely on an aspect of water governance as well as on concepts and measures of network analysis, a real comparative analysis among cases is impossible, and thus cannot be presented in the conclusions of this book. To compare these cases in the literal sense would result in a category error, as they differ—as discussed above—in terms of the level of governance, the types of networks, the sources behind network data, and other elements. This is why this book presents the broad possibilities and opportunities that Social Network Analysis offers for the analysis of diverse water governance issues and related research questions without restricting its scope to one single research design or methodological mode. This book thus broadly presents the different aspects of network concepts and measures when applied to study a given policy sector, and it aims to inspire future research to clearly define networks and related concepts for comparative analyses. Such approaches are important in order to be able to shed light not only on the core questions around who is central in networks and what influences network ties, but also for the broader questions around efficiency, quality, and legitimacy, that we still have much to learn about.

Structure of the Book

According to the logic laid out in this introduction, the book is structured as follows. After this introduction, a conceptual chapter discusses the literature and the most important developments, concepts, and questions in water governance. It then connects water governance elements to aspects of networks, and presents network concepts and measures that could be (and are) used in order to study these aspects. The heart of the book follows the conceptual chapters, and is made of nine case studies chapters. The first three chapters by Robbins and Lubell, Hollway, and Angst and Fischer study cases where network structures mainly highlight fragmentation of and clustering within networks. The three subsequent chapters by Mancilla and Bodin, Koebele et al., and Herzog and Ingold address how network fragmentation in water politics can potentially be overcome. More concretely, enhanced information exchange in forums, collaboration among coalition peers, or collaborative institutions are presented as key elements to reduce network fragmentation. The same holds true for single actors playing a key role in policy networks, an element emphasized in the last three case study chapters by Ebrahimiazarkharan et al., Olivier et al., and Bell and Henry. More specifically, these chapters emphasize different types of centralities in different types of networks. The concluding chapter summarizes the findings, synthesizes the broader implications for science and practice, and presents questions for a future research agenda related to networks in water governance.

References

Angst, M. (2018). Bottom-Up Identification of Subsystems in Complex Governance Systems. *Policy Studies Journal.* Published online first: 29 November 2018. https://doi.org/10.1111/psj.12301.

Berardo, R., & Lubell, M. (2019). The Ecology of Games as a Theory of Polycentricity: Recent Advances and Future Challenges. *Policy Studies Journal, 47*(1), 6–26.

Berardo, R., & Scholz, J. T. (2010). Self-Organizing Policy Networks: Risk, Partner Selection, and Cooperation in Estuaries. *American Journal of Political Science, 54*(3), 632–649.

Bodin, Ö. (2017). Collaborative Environmental Governance: Achieving Collective Action in Social-ecological Systems. *Science, 357*(6352), 1114.

Bodin, O., & Prell, C. (2011). *Social Networks and Natural Resource Management: Uncovering the Social Fabric of Environmental Governance.* Cambridge, UK: Cambridge University Press.

Feiock, R. C., & Scholz, J. T. (2009). *Self-organizing Federalism: Collaborative Mechanisms to Mitigate Institutional Collective Action Dilemmas.* Cambridge, UK: Cambridge University Press.

Galaz, V. (2007). Water Governance, Resilience and Global Environmental Change—A Reassessment of Integrated Water Resources Management (IWRM). *Water Science and Technology, 56*(4), 1–9.

Ingold, K., Fischer, M., de Boer, C., & Mollinga, P. P. (2016). Water Management Across Borders, Scales and Sectors: Recent Developments and Future Challenges in Water Policy Analysis. *Environmental Policy and Governance, 26*(4), 223–228.

Jaramillo, F., Desormeaux, A., Hedlund, J., Jawitz, J. W., Clerici, N., Piemontese, L., & Celi, J. (2019). Priorities and Interactions of Sustainable Development Goals (SDGs) with Focus on Wetlands. *Water, 11*(3), 619.

Lubell, M. (2013). Governing Institutional Complexity: The Ecology of Games Framework. *Policy Studies Journal, 41*(3), 537–559.

Lubell, M., & Edelenbos, J. (2013). Integrated Water Resources Management: A Comparative Laboratory for Water Governance. *International Journal of Water Governance, 1*, 177–196.

Maag, S., & Fischer, M. (2018). Why Government, Interest Groups, and Research Coordinate: The Different Purposes of Forums. *Society and natural resources, 31*(11), 1248–1265.

Pahl-Wostl, C., Holtz, G., Kastens, B., & Knieper, C. (2010). Analyzing Complex Water Governance Regimes: The Management and Transition Framework. *Environmental Science and Policy, 13*(7), 571–581.

Sabatier, P. A., & Weible, C. M. (2014). *Theories of the Policy Process*. Boulder, CO: Westview Press.

Weible, C. M., Heikkila, T., Ingold, K., & Fischer, M. (2016). *Policy Debates on Hydraulic Fracturing: Comparing Coalition Politics in North America and Europe*. New York: Springer.

Weible, C. M., & Sabatier, P. A. (2005). Comparing Policy Networks: Marine Protected Areas in California. *Policy Studies Journal, 33*(2), 181–201.

2

Conceptual Reflections About Water, Governance, and Networks

Manuel Fischer and Karin Ingold

Introduction

This chapter lays out the conceptual foundations of this book. We start with defining water governance and the different aspects that come into play when governing, steering, using, or managing the resource that is water. We then bring specificities of water and the governance of it as a resource together with key concepts and measures of social network analysis (SNA). To give an example, water is a multi-facetted topic with many issues such as hydrology, flood prevention, urban water management, irrigation, and so on. On top of this, water also relates to many other sectors such as agriculture, biodiversity, and climate change. These multi-issue interactions can be conceptualized as a network where several issues

M. Fischer (✉) • K. Ingold
Federal Institute of Aquatic Science and Technology, Eawag,
Dübendorf, Switzerland

University of Bern, Bern, Switzerland
e-mail: manuel.fischer@eawag.ch; karin.ingold@ipw.unibe.ch

© The Author(s) 2020
M. Fischer, K. Ingold (eds.), *Networks in Water Governance*, Palgrave Studies in Water Governance: Policy and Practice, https://doi.org/10.1007/978-3-030-46769-2_2

are more or less strongly linked to each other, depending upon, for instance, how many legal texts regulate two issues simultaneously, whether both issues are relevant for the same Sustainable Development Goals (SDGs), or whether biophysical links along a water stream create dependencies between two geographical areas. This is just one simple example of a one-mode issue network related to water and water management (see also Angst and Fischer; Koebele et al., this book).

Governing such complex issues requires interactions among a multitude of interested and concerned actors, representing different levels or sectors. This multitude of actors and their interactions can be represented as a network, and the phenomenon has been conceptualized accordingly through, for instance, collaborative governance, co-management, network governance, polycentric or adaptive governance (Ansell and Gash 2008; Emerson et al. 2012; Fischer and Leifeld 2015; Berardo and Lubell 2016). Collaborative networks among actors are assumed to create trust and capacities for learning, reduce conflicts, and consequently improve the quality of outcomes (e.g., Ansell and Gash 2008; Emerson et al. 2012). Network concepts and measures have been frequently used for studying structures of collaborative governance (Lubell et al. 2010; Berardo et al. 2014; Ulibarri and Scott 2017; Berardo et al. 2020). Networks of actors can be a reality that one can describe in water governance (see Herzog and Ingold or Bell and Henry, this book). At the same time, approaches of collaborative and network governance are often prescribed by principles of sustainability (e.g., SDG 17 on participatory governance, or examples provided by Ebrahimiazarkharan et al., this book), by rules and protocols (the European Union Water Framework Directive EUWFD, or examples presented by Olivier et al., this book), or fostered by forums and other collaborative institutions (Fischer and Leifeld 2015; Maag and Fischer 2018; Mancilla and Bodin, this book).

This conceptual chapter links characteristics of water governance to network concepts and measures. It does so by reviewing the most important literature in both fields—water governance and social network analysis—and by previewing how subsequent chapters use the respective concepts and measures to study their water governance case. The remainder of this chapter is structured as follows. First we discuss the concept of water governance, the related challenges, and approaches of network governance, collaborative governance, or ecosystem governance. Then we segue from water governance to networks and discuss what a network in water governance can be, that is, what can constitute different network nodes and ties and related sources of information. We then go on to discuss theoretical aspects of different types of network concepts and measures and their usefulness in structuring the empirical analysis of water governance. The last part of the chapter focuses exclusively on networks, in that it presents measures of network analysis and examples of what questions these could answer. It is here that we also discuss statistical approaches to network analysis and related research questions, and we finish with a synthesis of the main arguments pertaining to the effects of different network structures as well as common shortcomings of applied network analyses.

Water Governance: Challenges and Approaches to Deal with Them

Challenges of Dynamic Interdependencies in Water Governance

Solving water-related problems can be a difficult task. Governing waters means governing a complex natural ecosystem impacted by drivers from other ecological systems, and most importantly societal drivers through water use and related technical interventions. There exist different attempts at fostering the coordination between different stakeholders and jurisdictions using or being affected by the same resource: water (Wolf et al. 2003; Earle et al. 2011; Berardo and Lubell 2016). In this book, we aim to illustrate how network analysis concepts and measures can help grasp the fragmented and complex nature of water governance. We further show where and how coordination, cross-sectoral management, and multi-level collaboration is happening, and discuss related challenges and ways of improving the governance of water and making it fit for important interrelated current and future challenges related to climate change, growing populations, new types of pollutants, or biodiversity loss. Based on the discussion above, we focus on two important challenges related to water governance. When discussing network concepts and measures later in this chapter, we outline how these elements could help address and potentially overcome these challenges.

First Challenge: Water Issues Are Questioning Established Organization Within Borders, Sectors, and Levels

Three types of structures of governance systems complicate a holistic approach to water governance. First, water is a complex and multi-functional resource. As such, it embraces no less than four ecosystem service functions (Costanza et al. 1997): a provisioning function such as drinking water, a recreational function such as different water sports, a supporting function, for example, for soil formation, and a regulating function, for example, for substance dilution or nutrient buffering. Many parts of society thus depend upon water: its quality, quantity, abundance, and accessibility (Daily 1997). Water as a resource often also depends on and influences other resources such as land, energy, or biodiversity (Galaz 2007; Ingold et al. 2016; Lubell and Edelenbos 2013). It therefore connects the three core pillars (use of, protection of, protection from) to a multitude of other sectors such as agriculture, industry, tourism, education, or spatial planning, among others (Jønch-Clausen 2004; Halbe et al. 2013; Angst 2020). Yet, on the level of public administrations, as well as in the logic of many other actors, these sectoral aspects are dealt with in different units that interact only partially. Applying an integrated view to the regulation of the resource water is a challenge for policymakers (Hering and Ingold 2012).

Second, the logic of political borders and boundaries has seldom followed the rules of nature: jurisdictions and legal units rarely match the area of physical, chemical, ecological, or geological extent of a certain problem (Varone et al. 2013); and the implications of this fact are amplified if a problem transcends those political or legal borders (Ingold et al.

2018). In large surface waters (see Ebrahimiazarkharan et al.; Herzog and Ingold, this book), or where water quality or quantity is dependent from sources laying further away (Olivier et al., this book), coordination is complicated by administrative borders, and upstream activities might heavily affect or even hamper downstream activities related to the resource water in question. The existence of administrative boundaries complicates collaboration, as actors active in different administrative units have different logics, goals, and dynamics motivating them (Edelenbos and Van Meerkerk 2015; Treml et al. 2015; Ingold et al. 2016).

Third, governance systems are organized across different levels, from local to international. Including actors from different levels of governance can be particularly crucial in federalist settings (Hooghe and Marks 2001; Lubell 2013; Lubell and Edelenbos 2013). If competences are distributed at different levels of political organization—as is most often the case in water governance—we observe multi-level governance structures that create a need to collaborate across levels of governance (Hooghe and Marks 2001; Ingold 2014). Potential challenges arising in a multi-level system could include top-down initiatives colliding with bottom-up, voluntary, and self-organizing mechanisms (Berardo and Scholz 2010; Feiock and Scholz 2010). Another related challenge pertains to the relevant actors from other levels being included in policy making on a given level, such as local regions in European Union decision-making, or distant actors that have an influence on local systems through telecoupling mechanisms (Deines et al. 2016; Andriamihaja et al. 2019).

All three types of borders established by the organization of governance systems—between sectors and subsystems, between geographical-administrative units, and between levels of governance—point to potential misfit between social and ecological scales, that is, between the structure of the governance system supposed to deal with water issues, and the structure of the water issue itself. Social-ecological misfit is said to hamper the attainment of long-term sustainability goals and effective policymaking, implementation, and outcomes (Vignola et al. 2013; Guerrero et al. 2015; Treml et al. 2015; Bodin et al. 2019).

Second Challenge: Integrating Broad Ranges of Interests and Expertise

Related to the fact that water issues concern many different sectors and subsystems, and often transcend administrative boundaries, integrating a diversity of actors with different interests and expertise is a second core challenge in water governance (Ingold et al. 2018). What is already covered by the first challenge is the need to integrate actors from across sectoral and administrative borders, and across different levels (see also Ingold et al. 2016, 2018). The discussion here focuses on different types of actors in terms of the societal sectors they represent, the related interests and expertise they bring to the table, and the different goals and priorities, organizational structures, and professional languages these actors might have (Huxham et al. 2000). A simple three-layered categorization of actors can help in assessing the broad range of interests and expertise (Edelenbos et al. 2011; Crona and Parker 2012; Maag and Fischer 2018), but other and more fine-grained categorizations are, of course, possible.

First, state actors such as government and public administration have the task to develop, set, and implement binding rules (Crona and Parker 2012). Second, interest groups and non-governmental organizations (NGO) organize and represent different types of private and public interests in society, and feed these interests into the policy making process (Crona and Parker 2012). These actors also include trade associations, professional associations, civil society groups, as well as individual firms and private persons representing their interests. Whereas these two actor types are important for any policy sector, integrating actors representing science and research is especially crucial in water governance, which is grounded in social-ecological interdependencies (Galaz et al. 2008; McGinnis and Ostrom 2014). For example, in order to find appropriate governance solutions to water or environmental problems, the exact sources and effects of environmental problems need to be known. From social-hydrological modeling we know that causation, prevalence, temporality, and the impacts of complex phenomena such as climate change, nuclear waste storage, or toxic chemicals in waters, constantly interact with human and societal dimensions, which calls for a coupled human-water systems perspective (Sivapalan et al. 2012; Chen et al. 2016). In order to deal with the biophysical complexities of water, it seems thus crucial to integrate actors with scientific expertise and from practice: but the design and implementation of effective and efficient science-policy-practice interfaces still seems difficult and this for various reasons (van Enst et al. 2014; Raadgever et al. 2011; Brugnach et al. 2008; Sarewitz and Pielke 2007; Meadowcroft 2009)

Water Policies and Regulation

From Early Development Until Today

When water policies and regulations first started to develop, they were traditionally designed as reactions to specific individual problems: Regions regularly hit by flood events such as coastal areas or mountains started to establish flood prevention and risk management as early as the nineteenth century (Weingartner et al. 2003). State intervention was restricted mainly to infrastructural measures such as dikes, dams, and riverbed corrections (Ek et al. 2016). In other areas of the world, events and crisis related to water quality issues and pollution were major impetus for the development of water policies. Policymaking reacted to industrial accidents or sewage problems that lead to phenomena such as fish kill or the spread of diseases. Consequently, most western countries established environmental regulation in general, and water quality and urban water management in particular, between the 1950s, 1960s, and 1970s of the last century (Driessen and Glasbergen 2001). In sum, water policymaking was often simply the reaction to major events and crisis, designed in punctual, sectoral, and non-integrated manner.

This history results in today's multitude of sectoral and sub-sectoral policies, laws, acts, and strategies separately regulating transport on waters, water quality and quantity, flood and droughts, aquatic ecosystems, hydropower, and many more ecosystem functions provided by water. Some of these policies and policy sectors fall under the responsibility of the same administrative agency, while others involve many different authorities and agencies. Similarly, a broad range of diverse non-state actors also struggle to integrate the different water-related issues and challenges into their portfolio in a coordinated manner.

An analogous picture emerges on the international level (see Hollway, this book): no UN framework convention comparable to those for climate change or biodiversity addresses the protection of the water resource as a whole (see also Gleick 1998). Under the umbrella of the United Nations Economic Commission for Europe (UN-ECE), there exist regional water agreements, but the global International Water Courses Convention launched in 1997 only entered into force in 2014, and gets only limited participation, visibility and, outreach compared to other UN treaties. Most generally and recently, the Sustainable Development Goal 6 (SDG; Pärli and Fischer 2020) deals with water. It is committed to "Ensuring availability and sustainable management of water and sanitation for all" (United Nations 2015) and is composed of six substantive targets related to drinking water access, sanitation, water quality, efficient water use, integrated water management, and healthy water-related ecosystems. Additionally, SDG 14 deals with oceans and emphasizes the reduction of marine pollution, the restoration of ecosystems and conservation of coastal and marine areas, the reduction of acidification, or sustainable fishing, and the use of marine resources. Particular, and often discussed, challenges related to SDG implementation lie in the implementation across the global North and South, and in the complex trade-offs and synergies between different SDGs (Nilsson et al. 2016; Weitz et al. 2018).

Integrated Water Resource Management

Water governance developed over the last decades has typically tried to respond to the different entangled problems, sectors, and challenges of this resource (Rogers and Hall 2003; Gain and Schwab 2012). On the more applied side, the principle of Integrated Water Resources Management (IWRM) aims to address this multi-facetted nature of water, and substantively relies on three core pillars: the use of, the protection of, and the protection from water. Further, initiatives of Environmental

Policy Integration, or Whole-of-Government approaches (Jordan and Lenschow 2010; Tosun and Lang 2017) are among the most prominent and widely discussed solutions for organizing the complexity of water and environmental issues. These approaches aim to enhance cross-sectoral coordination and including environmental objectives in different sectoral logics in order to minimize procedural or substantive contradictions (Lafferty and Hovden 2003).

In Europe, transboundary and integrated water body management is specifically encouraged by the European Water Framework Directive (EUWFD; Kallis and Butler 2001; Earle et al. 2011). The EUWFD—currently under revision—aims at establishing the central idea of an ecosystem-based water governance at river basin scales, an approach that challenges the established organization of hierarchically organized silos in public administrations and structures of borders across states, regions, and local governments. Stakeholder and public participation are other main principles of the EUWFD. Furthermore, and again in line with main challenges of water governance as laid out in this book, the EUWFD requires actions within other policy fields (e.g., infrastructure, urban planning, agriculture, energy). Thus, there exist some efforts on the international level in joining forces and setting up transboundary initiatives of integrated water management, within, and beyond the EUWFD. All over the world, different river basin and catchment area associations exist in domestic or transboundary settings (e.g., Garrick et al. 2016; Jager 2016; Metz and Fischer 2016; see also Herzog and Ingold, Mancilla and Bodin, this book).

Collaborative Environmental Governance

On the more academic side, one approach to tackle these complexities and to bridge fragmented social-ecological systems is through collaborative governance (Ansell and Gash 2008), that is, the steering of an issue through the inclusion of a variety of public and private actors, all interested in finding a solution to the problem (Driessen et al. 2012). Such collaborative approaches are prominent in the literature on environmental and ecosystem governance, as well as in "sustainability science" in general (Lubell 2003; Koppenjan and Klijn 2004; Carlsson and Berkes 2005). They bring public and private stakeholders together in collective forums to engage in consensus-oriented decision-making (Ansell and Gash 2008). Such deliberative, bottom-up ways of collective problem-solving are said to improve environmental outcomes (Berkes and Folke 2002; Christensen et al. 2012; Newig 2012).

However, there is often a divergence between the claim that all relevant actors should collaborate and the empirical reality of who in fact works together. Actors do not only work together according to claims of inclusive, collective problem-solving, but they interact because of a variety of reasons (Calanni et al. 2014; see also Robbins and Lubell, this book). Networks of collaborative governance are shaped by institutions, power, resources, tasks, problem perceptions, or preferences of actors (Berardo and Scholz 2010; Lubell et al. 2010; Henry 2011; Leifeld and Schneider 2012; Ingold 2011; Ingold and Fischer 2014; Fischer and Sciarini 2016). One of the most prominent hypotheses in SNA is the homophily hypothesis: "birds of the same feathers flock together" (McPherson et al. 2001): actors having the same convictions, being similarly affected by a problem, playing the same role in solution-finding, and so on have the tendency to interact with each other. Different chapters in this book also show that actors who think alike, who share similar beliefs, and who advocate for the same policy solutions have the tendency to coordinate actions together (see Mancilla and Bodin, Koebele et al., this book). Thus, water governance is probably no different than other policy sectors in that respect, and is characterized through power games and authority. Thus, one way to foster coordination in water governance could be overcoming ideological conflict (applicable across different levels, including the international one, see Hollway, this book).

Another salient concept that emerges in both the literature on environmental collaborative governance as on networks is that of social capital and resilience, which should facilitate coordination across actors and thus make the entire network stronger. In times of shocks and catastrophes, such as flood events or water scarcity, enhanced social capital and resilience might reduce the vulnerability of affected communities. Different chapters in this book address the issue of social capital and how to enhance cohesion or reduce fragmentation in networks (Hollway;

Mancilla and Bodin; Herzog and Ingold; Robbins and Lubell; Koebele et al.; Ebrahimiazarkharan et al.; Olivier et al., this book). Olivier et al. (this book) investigate relationships in the medium of communication and relate them back to prescribed rules that can be found in protocols. Behavioral expectations can not only be linked to rules and formal institutions, but also to trust and reciprocity. Frequent meetings, participating in the same venues and forums, and being members of the same associations seem to decisively shape water governance arrangements (Lubell and Edelenbos 2013; Herzog and Ingold 2019). The chapters by Mancilla and Bodin as well as Robbins and Lubell (this book) explicitly analyze the co-participation in forums as an important driver for coordination.

Most of the contributions in this book are addressing in some way or the other the question of how to overcome fragmentation (be it across sectors, levels, jurisdictions, space, time, or scales) and how to foster coordination and collaboration among actors, given the complexities of water governance and the related risks of fragmentation. Besides similar roles or ideologies, the co-participation in forums or prescribed rules enhancing interactions, single key actors can make a network into a so-called complete graph. Without bridging actors, the whole network could fall apart into several sub-networks, sub-graphs, or sub-sections. The role of such key actors is explicitly addressed in the chapters by Bell and Henry, Angst and Fischer, but also Ebrahimiazarkharan et al. and Herzog and Ingold (this book) identify single actors with high network centralities.

Networks and Network Analysis

Social network analysis (SNA) is a method of detecting and interpreting structures and patterns of connections between actors who may be individuals, collectives, or institutions. SNA has been increasingly used in a variety of fields from political science (Fischer and Sciarini 2016; Victor et al. 2017) to business marketing (Iacobucci 1996), or in general in many different fields of the social sciences (Borgatti et al. 2009). SNA is a versatile tool for different applications due to its graphical representation, structural intuition, and systematic data interpretation (Freeman

2004). More substantively, SNA is designed to deal with data on relations among entities, and thus data that describes interconnected phenomena and consists of non-interdependent observations. Whenever relations among entities are crucial for understanding a given phenomenon, SNA can provide important insights. Thus, adopting a network approach is one prominent way to study interrelated, multi-level, and cross-sectoral governance arrangements (Berardo and Scholz 2010; Lubell 2013). Network approaches are used for both understanding and describing water governance arrangements, as well as for providing more normative or practice-oriented recommendations.

Nodes and Ties in Water Governance Networks

Network Nodes

The two defining elements of networks are a set of nodes that are connected through ties (other authors add flows and protocols to networks, see e.g., Ulibarri and Scott 2017). Nodes in governance networks are social actors such as individuals, organizations, or political-administrative entities, or a mix of all of the above (Huxham et al. 2000). Organizational actors can be interest groups, political parties, administrative units, research centers, or other actors involved in governance processes, while individuals can be representatives of these organizations. For example, in this book, several chapters analyze organizational actors (Angst and Fischer, Herzog and Ingold, Robins and Lubell, Bell and Henry). Other chapters focus on individual actors, as we see in Mancilla and Bodin's study of the relations among individuals participating in the river basin forum. Yet others deal with a mix of organizational and individual actors (Koebele et al., this book). Finally, political-administrative entities such as states (Hollway, this book) or sub-states (Fischer and Jager 2020) can be nodes in networks of water governance.

Besides social actors as nodes, the interdisciplinary and flexible nature of the network approach also allows to integrate other types of nodes besides social actors, for example to take into account social-ecological interdependencies or relations between social actors and issues or institutions in water governance. The concept of the ecology of games (Lubell 2013) describes forums as network nodes that actors participating in these forums are related to. In a similar vein, the social-ecological network approach connects social actors to a second type of node, that is, ecological elements such as forest patches, wetlands, or river branches (Bodin and Tengö 2012; Bodin 2017; Ingold et al. 2018). Furthermore, governance issues or policy problems have been defined as network nodes that actors are connected to (Angst 2020; Brandenberger et al. 2020). Angst and Fischer (this book) analyze ties between actors and 26 different issues in Swiss water politics, whereas Bell and Henry (this book) focus on actors and their participation in 526 events. Koebele et al. (this book) link actors to policy beliefs and positions, and Olivier et al. (this book) analyze actors being linked to formal rules and protocols.

Network Ties

Network ties between social actors are usually ties of coordination, information exchange, venue co-participation, conflict, or formal cooperation based on contracts. In this book, chapters deal with network ties between two nodes that go from collaboration or coordination among actors (Herzog and Ingold, Mancilla and Bodin, Hollway, Koebele et al.), to influence attribution among actors (Mancilla and Bodin), to shared believes among actors (Mancilla and Bodin, Koebele et al.), to joint responsibility for an issue due to formal laws (Olivier et al.), to conflict (Hollway), to event co-participation (Bell and Henry).

Of course, there is always some simplification involved in the definition of a tie between network nodes. In reality, social relations can be very complex. For example, Ansell and Gash (2008) define five steps of the collaborative process (dialogue, trust, commitment, shared understanding, outcomes). Thus, ties can not only be present or absent, but they can have different intensities and weights, such as different intensities of collaboration (Margerum 2008). Furthermore, cooperation and conflict, for example, can also exist in parallel, and be assessed in so-called multiplex networks (see Koebele et al., this book). Thus, social interactions can exist on a variety of planes, and each realization of a social interaction aggregates to form a more or less discrete relation. Furthermore, some of these multiplex ties influence each other, as when venue co-participation leads to more intense exchanges of information (Leifeld and Schneider 2012; Fischer and Sciarini 2016). For example, the chapter by Mancilla and Bodin takes into account three parallel relations between actors (coordination, influence attribution, shared beliefs) and analyze how forum co-participation influences any of these relations. Koebele et al. (this book) also define governance as the existence of multiple parallel types of actor interactions and compare the networks of joint policy beliefs, joint policy positions, and coordination among actors.

Nodes and ties can combine differently and give rise to different types of networks (see Wassermann and Faust 1994). One-mode networks are the simplest form of networks, as they involve one type of node and (usually) one type of ties between these nodes. Example of one-mode networks in this book can be found in the chapters by Hollway, Herzog and Ingold, Mancilla and Bodin, or Ebrahimiazarkharan et al. More complex types of networks are two-mode or bipartite networks where two different types of nodes are connected (without being connected among themselves). Examples in this book are actors dealing with issues (Angst and Fischer), actors participating in events (Bell and Henry), actors adhering to rules (Olivier et al.), or actors sharing policy beliefs and positions (Koebele et al.). In the case of the chapter by Bell and Henry, the two-mode network is then transformed into a one-mode network in order to represent actor coordination through joint participation in events. Olivier et al. discuss their two-mode network as a hypergraph, composed of hyperedges, that is, rules that can link to many different actors. Finally, multi-level networks are networks with different types of nodes that are related to each other, and that include network relations within the different sets of nodes. There is no example of such a network in our book, but a new strand of literature also relevant to water governance conceptualizes social-ecological systems as two interrelated networks (Bodin 2017; Sayles et al. 2019; Bodin et al. 2019): a social network where actors are related to each other through collaboration; and ecological network where ecosystem units (like forest patches or species) are related to each other through interactions like species migration or pollination. Finally, both systems, the social and the ecological, are also linked to each other through mostly human interaction with the environment such as land use, land or resources management, or policy regulation. A claim by the social-ecological network literature proposes that for efficient and effective management of natural resources in general, and water in particular, there should be an alignment between the social and the ecological units and interactions (Bodin and Tengö 2012; Widmer et al. 2019).

Key Variables in Network Analysis

Networks can be analyzed on three different levels, the micro-level of individual actors in the network, the meso-level substructures within the network, that is, sets of nodes and ties in the network, or at the macro-level of the entire network (see also Wassermann and Faust 1994; Borgatti et al. 2018). The following discussion of the most prominent variables used in network analyses—in general and in the chapters of this book more specifically—build on this three-fold categorization. In this book, the chapters by Herzog and Ingold or Ebrahimiazarkharan et al. also rely on this three-layered categorization of network indicators. Readers interested in mathematical definitions and more thorough discussions of measures should consult the respective specialized literature (e.g., Wassermann and Faust 1994; Borgatti et al. 2018).

Micro-Level Network Measures: Centralities

The most common, important, and straightforward measure related to networks at the micro-level of network nodes is centrality. Most generally, centrality describes how central an actor is within the network, that is, the relational position of a given node in the overall network. Centrality, however, can take different forms (Freeman 1978; Friedkin 1991), although they are often strongly correlated (Valente et al. 2008). First, degree centrality is based on the simple number of incoming (in-degree) and outgoing ties (out-degree) of a node. Second, betweenness centrality takes into account the degree to which a node is located on the shortest path between any two other nodes in the network. Network nodes with high betweenness centralities are potentially important bridging actors or brokers in the network. Third, closeness centrality identifies the actors with the shortest paths (sequence of ties) to all other actors in the network, on average. Herzog and Ingold (this book) rely on betweenness centrality to identify actors that can play bridging roles in their three local Rhine river catchments. Bell and Henry (this book) assess betweenness centrality as well as degree centrality and contrast it with qualitative evidence of the actors' brokerage and leadership activities. Ebrahimiazarkharan et al. (this book) rely on both degree centrality and betweenness centrality to identify important actors in watersheds. Finally, Olivier et al. (this book) assess betweenness and closeness centralities of rules in order to compare different types of rules.

Other measures of centrality exist, among others Eigenvector centrality that is based on degree centrality (Bonacich 2007). The measure counts how many other nodes a given node is connected to. It thus takes into account the centrality of the nodes a given node is related to by giving more weight to more central nodes. The concept of centrality further extends to two-mode networks (Opsahl et al. 2010), as is illustrated by Angst and Fischer (this book). They apply centrality measures adopted from ecology studies that calculated the centrality of actors in terms of their ties to issues within as well as beyond their subsystem. Koebele et al. (this book) compare centralities of actors in different one- and two-mode networks, and Olivier et al. (this book) assess in- and out-degree of both modes in actor—rule networks.

Meso-Level Network Measures: Clustering, Cliques, and Modularity

Clustering is a normal mechanism in networks: some types of actors will more strongly interact with some actors than with others (see Robbins and Lubell, this book). This results in the network being structured to some degree by denser clusters of nodes and less dense connections across these clusters. One way to identify clusters would correspond to assessing clusters that are different (denser) than the rest of the network, and there are different thresholds that one can set for identification purposes (see discussion in Angst and Fischer, this book). Angst and Fischer (this book) rely on a modularity algorithm to identify clusters in a two-mode network, and Robbins and Lubell (this book) use a walktrap community detection algorithm in their collaboration network. Another way of identifying clusters would correspond to pre-defining what the clusters should look like structurally. For example, one could define clusters as cliques, that is, a set of nodes in which every node is connected to every other node. Similarly, a faction is defined as a cohesive group of nodes whose number is pre-defined, and who need to be linked to each other (Borgatti et al. 2018; Wassermann and Faust 1994). Herzog and Ingold (this book) use faction analysis to identify coalitions of actors in their three case study areas along the Rhine river.

Another aspect of clustering is that it can be compared to actors' attributes. Robbins and Lubell (this book) discuss the different mechanisms potentially leading to clustering, that is, geographical proximity, shared policy beliefs, shared social group identification, shared sectoral affiliations, and trust and reciprocity among actors. Because actors tend to connect to those that are similar to them with respect to these attributes (homophily), the network ends up having a cluster structure. One can then assess to what degree these attributes correlate with the cluster structure of the network, for example by applying the E-I indicator that assesses to what degree nodes have ties within a given group or across groups (see Ebrahimiazarkharan et al. and Robbins and Lubell, this book), or through statistical models (see below).

Macro-Level Network Measures: Density, Centralization, Reciprocity, Transitivity, Core-Periphery

Measures to describe the structural properties of a network at the macro-level of the entire network abound, and most of them correspond to some kind of aggregate or average measure of micro- or meso-level structures. The most prominent example is the following. First, network density corresponds to the proportion of observed ties as compared to all theoretically possible ties in the network. The measure thus assesses how many network ties there are based on the overall number of possible ties.

Herzog and Ingold (this book) assess network density for measuring the cohesion of their network, and compare these measures across three networks. Ebrahimiazarkharan et al. rely on density, transitivity, and reciprocity parameters (see below) to describe basic properties of their network. Second, centralization of the network is an indicator to describe connectivity in the network (see Olivier et al., this book). Based on potentially any centrality measure (see above), centralization assesses to what degree centrality measures in a network are equally distributed among nodes. The measure—if high—thus potentially also indicates whether there is a hierarchy in the network in terms of centralities, with few central actors and many less central actors.

Third, reciprocity is actually a measure at the dyadic level (see below in relation to statistical model of networks), but is often represented as an average degree across the entire network. Reciprocity indicates what proportion of ties among two actors are reciprocated, that is, if node i indicates a tie to node j, node j also indicates a tie to node i. This measure is only valid for directed networks. Herzog and Ingold (this book) rely on this measure as one piece of evidence for assessing the cohesion of their network. Fourth, transitivity is similar to reciprocity but refers to triangular structures, that is, structures where nodes i and j are connected, they both also connect to node k.

Fifth, a core-periphery structure of a network is present if there is a set of nodes in the core of the network that are strongly inter-connected, and a set of nodes in the periphery of the network that are weakly related to nodes of the core, and not connected among themselves (Borgatti and Everett 2000). This represents an ideal-typical network structure to which empirically observed networks can be compared to. A core-periphery measure is used by Ebrahimiazarkharan et al. (this book) for the identification of core actors that might play a strong role in water management.

Statistical Models of Network Data

The measures presented above all point to descriptive aspects of network analysis. An inferential approach to network data—for example, analyzing network ties as dependent variables influenced by a set of independent variables—is complicated by the fact that network ties cannot be assumed to be independent from each other. Interdependencies reflect the very nature of network approaches and data: that it represents situations where different network nodes are interrelated. Standard regression models—that assume independency of their observations—would erroneously attribute explanatory power to actor attributes instead of endogenous network processes (Desmarais and Cranmer 2012; Cranmer et al. 2017). There are several options to overcome this problem with models that we discuss below.

Generally speaking, in all of these models, "independent variables" (or covariates) usually used to explain the structure of networks in such models can be categorized into three types (Leifeld and Schneider 2012; Lubell et al. 2012; Ingold and Fischer 2014). First, node attributes can be important as they can detect the specific activity or popularity of given types of actors. Second, attributes of relations between two actors (actor dyads) can matter. An example for dyad level substructures is the phenomenon of homophily. Homophily describes the effect that similar actors (from the same governance level, of the same organizational type, with similar specializations, etc.) tend to exchange information above average (McPherson et al. 2001; Calanni et al. 2014; Fischer and Sciarini 2016). Another example would be a tie of information exchange that (positively) influences a tie of support of some kind. Third, endogenous network structures independently of node or dyadic attributes matter, and describe the core issue of statistically modeling networks: the potential dependencies of the network structure on the network structure itself. Reciprocity (the phenomenon of "tit-for-tat", meaning an actor reciprocates a tie) and triadic closure (the phenomenon that can be summed up as "a friend of my friend tends also to be my friend") are typical endogenous network level factors.

Different statistical approaches to network data have appeared in the recent past. We only present a short overview related to the applications in this book. For more detailed discussions and tutorials, the interested reader should consult the relevant references. There are differences to these approaches. For example, dynamic models such as Stochastic Actor-Oriented Models (SAOMs) or Dynamic Network Actor Models (DyNAMs) can take into account the evolution of networks based on network observations at different time points, and thus also model the rate of network change (see Hollway, this book). Hollway (this book) also presents a short discussion of the differences among the different models for statistical network analysis. The approaches sometimes also differ in their epistemological and ontological foundations.

The chapter by Mancilla and Bodin in this book applies Quadratic Assignment Procedures (QAP). QAP allows for a comparison of several networks and thus assess a potential influence of one network tie on another. The method is based on correlational measures between two or more networks that are calculated by permutations of the networks, keeping network size and density constant (Dekker et al. 2007). As compared to the other network models below, QAP is unable to take endogenous network effects into account. Mancilla and Bodin (this book) use the method to assess correlations between the coordination and shared beliefs networks among actors.

Exponential Random Graph Models (ERGMs, Robins et al. 2007) allow modeling network structures by comparing them to large sets of random networks. ERGMs model the probability of observing a given configuration of the network, as compared to all other possible network configurations with the same number of nodes and network density (Cranmer and Desmarais 2011). In this book, Mancilla and Bodin apply an ERGM to their data on river management in Brazil in order to identify the factors that lead actors within a river basin forum to coordinate, beyond the correlational among networks they assess with QAP.

Stochastic Actor-Oriented Models (SAOMs; Snijders et al. 2010; Robbins and Lubell, this book) are able to take the evolution of networks over time into account, based on the observation of networks at specific points in time. Such models are actor-based in the sense that they model the individual choices of actors to create or delete ties at specific rates between two time points as a Markov process. SAOMs rely on network observations at different points in time, and the first observation serves as a baseline for modeling the evolution. In their chapter in this book, Robbins and Lubell rely on SAOMs to identify factors influencing the network of actors dealing with sustainable fisheries. Alternatively to SAOMs, but with a slightly different logic less dependent on assumptions about actors' behavior, Temporal ERGMs (TERGM) can also model the evolution of networks over time (Leifeld et al. 2018).

Dynamic Network Actor Models (DyNAMs) also deal with the evolution of networks, and enable the researcher to model the exact time stamps of given events such as tie creation. This means that the data analyzed by these models is not composed of network observations at given points in time, but by network ties that are created or deleted at continuous time points. DyNAMs simulate the most likely series of changes leading to a given network structure through an actor-oriented approach (with the same logic as SAOMs, Stadtfeld et al. 2017; Hollway, this book). Hollway (this book) uses DyNAMs in order to model the evolution of the network of international water treaties among nation states. Alternatively, Relational Event Models (REMs) have a similar goal than DyNAMs (Butts 2008; Brandenberger 2018; see also discussion by Hollway, this book).

Common Limitations of Network Analytic Approaches

As with any conceptual and methodological approach, concepts and measures from network analysis can well analyze given phenomena, but are rather blind to other aspects. Therefore, network analytical tools can—and often are—combined with other methods (i.e., mixed methods, Domínguez and Hollstein 2014; Tashakkori and Teddlie 2010). In this book, Mancilla and Bodin as well as Bell and Henry combine network analysis with evidence from qualitative interviews in order to either compare evidence from both approaches and validate the theoretical claims of network indicators or to complement information on the functioning of the entire network and the institutional surroundings of the related forum (respectively).

A common challenge with network analysis is the definition of network boundaries and the question whether it is necessary to know whether the whole network, or only parts of it, need to be known in order to draw relevant conclusions (Wasserman and Faust 1994). There are situations when the relevant actors forming the network are easy and unanimous to identify, as is the case with formal participants of a policy process (Fischer and Sciarini 2016; Ingold and Fischer 2014) or members of a forum (Fischer and Leifeld 2015; Mancilla and Bodin, this book). Yet, given the complexities of water-related policy problems that cross administrative or sectoral boundaries, network boundary definition is most often not a preparation ahead of a proper network analysis, but might only be one of the results of an explorative network analysis. For example, Angst and Fischer (this book) use a large network of actors and issues in Swiss water politics to identify boundaries of water-related sub-systems, or Koebele et al. (this book) identify coalitions of actors based—among others—on network data. Related to the issue of boundary definition and the knowledge of which actors belong to the network and which ones do not is the issue of missing data. As with any quantitative method, missing data on actors and ties in the network can lead to biased results (Kossinets 2006; Smith and Moody 2013; Berardo et al. 2020). Relying on grey literature (Bell and Henry, this book) or newspaper reports (Koebele et al., this book) can be a valid alternative to survey or interview data, but these sources might suffer from reporting biases (see discussions by Bell and Henry or Koebele et al., this book).

Another challenge is that networks evolve over time due to actors' strategic behavior. Actors create and abandon network ties in pursuit of their goals, constrained by costs and uncertainties, formal authority, existing network relations, and limited cognitive capacities (Scholz et al. 2008; Snijders et al. 2010). Furthermore, changing contextual factors can influence network structures. Water-related issues have the tendency to constantly evolve, change, and interact with other phenomena, and complex feedback-loops characterize water-related ecosystems (Vogel 2004; Pahl-Wostl et al. 2010; Crona and Parker 2012). Thus, the importance and the purpose of collaboration in networks might change as the collaborative process itself unfolds (Huxham et al. 2000, p. 345). Assessing network dynamics is a challenge for researchers, especially depending on the type of data-gathering process they rely on. Ample resources are needed to do interviews or surveys at different points in time, whereas coding document or newspaper data that report on network interactions at different points in time is more straightforward. In this book, Koebele et al. (this book) rely on data coded from newspaper articles between 2005 and 2014, but decide to analyze only aggregated versions of the network.

Considering the evolution of the network over time could be difficult when the information is based on newspaper reports, as the network at each point in time is likely quite sparse. Hollway (this book) as well as Robins and Lubell (this book) explicitly model network evolution over time, by relying on statistical models of DyNAMs and SOAMs, respectively, and data based on existing datasets on international water relations, and three waves of surveys, respectively.

Finally, following the literature on networks and collaborative governance, the network structure influences policy outputs and outcomes (Koppenjan and Klijn 2004; Weible and Sabatier 2005; Lubell et al. 2012). It is claimed that networks can impact the quality of natural resources management due to stakeholder involvement increasing information exchange, boundary organizations linking a variety of different actors, and knowledge transfers (Adger 2003; Crona and Bodin 2006; Duit and Galaz 2008; Prell et al. 2009; Crona and Parker 2012). Furthermore, in the approach of social-ecological networks, social-ecological fit is supposed to lead to good management outcomes and sustainable resource governance, also related to water resources (Bodin 2017). However, even if collaborative forms of governance hypothetically continued to spread, there is not yet consensus about their actual effectiveness or performance (Gerlak et al. 2013). Compared to the number of similar claims, there is to date only little empirical evidence on how networks influence outcomes. There are many different orders of outputs, ranging from the creation of collaborative networks themselves, to the negotiation of policy solutions, to longer-term processes such as learning (Bryson et al. 2006), or even to environmental and ecological consequences such as healthier ecosystems (Bodin et al. 2019). On a theoretical level, networks and their structures are only one factor that influences outputs and outcomes, and on a practical level, comparing different networks or analyzing them over time in order to approach some idea of causality is often complicated. In this book, we nevertheless address some of these challenges and hive some new pathways in how to study complex water governance systems across levels, countries, and via diverse social network applications.

References

Adger, W. N. (2003). Social Capital, Collective Action, and Adaptation to Climate Change. *Economic Geography, 79*(4), 387–404.

Andriamihaja, O. R., Metz, F., Zaehringer, J. G., Fischer, M., & Messerli, P. (2019). Land Competition Under Telecoupling: Distant Actors' Environmental Versus Economic Claims on Land in North-Eastern Madagascar. *Sustainability, 11*(3), 851.

Angst, M. (2020) Bottom-Up Identification of Subsystems in Complex Governance Systems. *Policy Studies Journal.* Online first.

Ansell, C., & Gash, A. (2008). Collaborative Governance in Theory and Practice. *Journal of Public Administration Research and Theory, 18*(4), 543–571.

Berardo, R., & Lubell, M. (2016). Understanding What Shapes a Polycentric Governance System. *Public Administration Review, 76*(5), 738–751.

Berardo, R., & Scholz, J. T. (2010). Self-Organizing Policy Networks: Risk, Partner Selection, and Cooperation in Estuaries. *American Journal of Political Science, 54*(3), 632–649.

Berardo, R., Heikkila, T., & Gerlak, A. K. (2014). Interorganizational Engagement in Collaborative Environmental Management: Evidence from the South Florida Ecosystem Restoration Task Force. *Journal of Public Administration Research and Theory, 24*(3), 697–719.

Berardo, R., Fischer, M., & Hamilton, M. (2020). *Collaborative Governance and the Challenges of Network-Based Research.* Submitted Manuscript.

Berkes, F., & Folke, C. (2002). Back to the Future: Ecosystem Dynamics and Local Knowledge. In L. H. Gunderson & S. H. Holling (Eds.), *Panarchy: Understanding Transformations in Human and Natural Systems* (pp. 131–146). Washington, DC: Island Press.

Bodin, Ö. (2017). Collaborative Environmental Governance: Achieving Collective Action in Social-Ecological Systems. *Science, 357*, eaan1114.

Bodin, Ö., & Tengö, M. (2012). Disentangling Intangible Social–Ecological Systems. *Global Environmental Change, 22*, 430–439.

Bodin, Ö., Alexander, S. M., Baggio, J., Barnes, M. L., Berardo, R., Cumming, G. S., Dee, L. E., Fischer, A. P., Fischer, M., Mancilla Garcia, M., Guerrero, A. M., Hileman, J., Ingold, K., Matous, P., Morrison, T. H., Nohrstedt, D., Pittman, J., Robins, G., & Sayles, J. S. (2019). Improving Network Approaches to the Study of Complex Social–Ecological Interdependencies. *Nature Sustainability, 2*, 551–559.

Bonacich, P. (2007). Some Unique Properties of Eigenvector Centrality. *Social Networks, 29*(4), 555–564.

Borgatti, S. P., & Everett, M. (2000). Models of Core/Periphery Structures in Networks. *Social Networks, 21*(4), 375–395.

Borgatti, S. P., Mehra, A., Brass, D. J., & Labianca, G. (2009). Network Analysis in the Social Sciences. *Science, 323*(5916), 892–895.

Borgatti, S. P., Everett, M. G., & Johnson, J. C. (2018). *Analyzing Social Networks*. London: Sage.

Brandenberger, L. (2018). Trading Favors – Examining the Temporal Dynamics of Reciprocity in Congressional Collaborations Using Relational Event Models. *Social Networks, 54*, 238–253.

Brandenberger, L., Ingold, K., Fischer, M., Schläpfer, I., & Leifeld, P. (2020). *Overlapping Network Structures: Why Actors Engage in Diverse Policy Issues*. Submitted manuscript.

Brugnach, M., Dewulf, A., Pahl-Wostl, C., & Taillieu, T. (2008). Toward a Relational Concept of Uncertainty: About Knowing Too Little, Knowing Too Differently, and Accepting Not to Know. *Ecology and Society, 13*(2), 30.

Bryson, J., Crosby, C. M. B., & Middleton Stone, M. (2006). The Design and Implementation of Cross-Sector Collaborations: Propositions from the Literature. *Public Administration Review, 20*(1), 45–55.

Butts, C. T. (2008). A Relational Event Framework for Social Action. *Sociological Methodology, 38*(1), 155–200.

Calanni, J. C., Siddiki, S. N., Weible, C. M., & Leach, W. D. (2014). Explaining Coordination in Collaborative Partnerships and Clarifying the Scope of the Belief Homophily Hypothesis. *Journal of Public Administration Research and Theory, 25*(3), 901–927.

Carlsson, L., & Berkes, F. (2005). Co-management: Concepts and Methodological Implications. *Journal of Environmental Management, 75*(1), 65–76.

Chen, X., Wang, D., Tian, F., & Sivapalan, M. (2016). From Channelization to Restoration: Sociohydrologic Modeling with Changing Community Preferences in the Kissimmee River Basin, Florida. *Water Resources Research, 52*(2), 1227–1244.

Christensen, P., Kornov, L., Holm Nielsen, E. (2012). Between Governance and Government: Danish EIA in Uncharted Waters. *Journal of Environmental Assessment Policy and Management, 14*(4). https://doi.org/10.1142/S1464333212500214.

Costanza, R., d'Arge, R., de Groot, R. S., Farber, S., Grasso, M., Hannon, B., Limburg, K., Naeem, S., O'Neill, R. V., Paruelo, J., Raskin, R. G., Sutton, P., & van den Belt, M. (1997). The Value of the World's Ecosystem Services and Natural Capital. *Nature, 387*, 253–260.

Cranmer, S. J., & Desmarais, B. A. (2011). Inferential Network Analysis with Exponential Random Graph Models. *Political Analysis, 19*(1), 66–86. https://doi.org/10.1093/pan/mpq037

Cranmer, S. J., Leifeld, P., McClurg, S. D., & Rolfe, M. (2017). Navigating the Range of Statistical Tools for Inferential Network Analysis. *American Journal of Political Science, 61*(1), 237–251.

Crona, B. I., & Bodin, Ö. (2006). What You Know Is Who You Know? Communication Patterns Among Resource Users as a Prerequisite for Co-Management. *Ecology and Society, 11*(2), 7.

Crona, B. I., & Parker, J. N. (2012). Learning in Support of Governance: Theories, Methods, and a Framework to Assess How Bridging Organizations Contribute to Adaptive Resource Governance. *Ecology and Society, 17*(1), 32.

Daily, G. C. (1997). *Nature's Services: Societal Dependence on Natural Ecosystems.* Washington, DC: Island Press.

Deines, J. M., Liu, X., & Liu, J. (2016). Telecoupling in Urban Water Systems: An Examination of Beijing's Imported Water Supply. *Water International, 41*(2), 251–270.

Dekker, D., Krackhardt, D., & Snijders, T. A. (2007). Sensitivity of MRQAP Tests to Collinearity and Autocorrelation Conditions. *Psychometrika, 72*(4), 563–581.

Desmarais, B. A., & Cranmer, S. J. (2012). Micro-level Interpretation of Exponential Random Graph Models with Application to Estuary Networks. *Policy Studies Journal, 40*(3), 402–434.

Domínguez, S., & Hollstein, B. (2014). *Mixed Methods Social Networks Research: Design and Applications* (Vol. 36). Cambridge: Cambridge University Press.

Driessen, P. P. J., & Glasbergen, P. (2001). New Directions in Environmental Politics: Concluding Remarks. In P. P. J. Driessen & P. Glasbergen (Eds.), *Greening Society; the Paradigm Shift in Dutch Environmental Politics* (pp. 245–262). Dordrecht: Kluwer Academic Publishers.

Driessen, P. P. J., Dieperink, C., van Laerhoven, F., Runhaar, H. A. C., & Vermeulen, W. J. V. (2012). Towards a Conceptual Framework for the Study of Shifts in Modes of Environmental Governance: Experiences from The Netherlands. *Environmental Policy and Governance, 22*(3), 143–160.

Duit, A., & Galaz, V. (2008). Governance and Complexity: Emerging Issues for Governance Theory. *Governance, 21*(3), 311–335.

Earle, J. R., Blacklocke, S., Bruen, M., Almeida, G., & Keating, D. (2011). Integrating the Implementation of the European Union Water Framework Directive and Floods Directive in Ireland. *Water Science and Technology, 64*(10), 2044–2051.

Edelenbos, J., & Van Meerkerk, I. (2015). Connective Capacity in Water Governance Practices: The Meaning of Trust and Boundary Spanning for Integrated Performance. *Current Opinion in Environmental Sustainability, 12*, 25–29.

Edelenbos, J., Van Buuren, A., & Van Schie, N. (2011). Co-producing Knowledge: Joint Knowledge Production Between Experts, Bureaucrats and Stakeholders in Dutch Water Management Projects. *Environmental Science and Policy, 14*(6), 675–684.

Ek, K., Pettersson, M., Alexander, M., Beyers, J.-C., Pardoe, J., Priest, S., Suykens, C., & van Rijswick, H. F. M. W. (2016). *Design Principles for Resilient, Efficient and Legitimate Flood Risk Governance – Lessons from Cross-Country Comparisons.* Utrecht: STAR-FLOOD Consortium.

Emerson, K., Nabatchi, T., & Balogh, S. (2012). An Integrative Framework for Collaborative Governance. *Journal of Public Administration Research and Theory, 22*(1), 1–29.

Feiock, R., & Scholz, J. T. (2010). *Self-Organizing Federalism: Collaborative Mechanisms to Mitigate Institutional Collective Action Dilemmas.* New York, NY: Cambridge University Press.

Fischer, M., & Jager, N. (2020). How Policy-Specific Factors Influence Horizontal Cooperation Among Subnational Governments: Evidence from the Swiss Water Sector. *Publius – The Journal of Federalism.* Online first. https://doi.org/10.1093/publius/pjaa002

Fischer, M., & Leifeld, P. (2015). Policy Forums: Why Do They Exist and What Are They Used For? *Policy Sciences, 48*(3), 363–382.

Fischer, M., & Sciarini, P. (2016). Drivers of Collaboration in Political Decision Making: A Cross-Sector Perspective. *The Journal of Politics, 78*(1), 63–74.

Freeman, L. (1978). Centrality in Social Networks: Conceptual Clarification. *Social Networks, 1*(3), 215–239.

Freeman, L. (2004). The Development of Social Network Analysis. *A Study in the Sociology of Science, 1*, 687.

Friedkin, N. (1991). Theoretical Foundations for Centrality Measures. *American Journal of Sociology, 96*(6), 1478–1504.

Gain, A. K., & Schwab, M. (2012). An Assessment of Water Governance Trends: The Case of Bangladesh. *Water Policy, 14*(5), 821–840.

Galaz, V. (2007). Water Governance, Resilience and Global Environmental Change – A Reassessment of Integrated Water Resources Management (IWRM). *Water Science and Technology, 56*(4), 1–9.

Galaz, V., Olsson, P., Hahn, T., & Svedin, U. (2008). The Problem of Fit Among Biophysical Systems, Environmental and Resource Regimes, and Broader Governance Systems: Insights and Emerging Challenges. In O. R. Young, H. Schroeder, & L. A. King (Eds.), *Institutions and Environmental Change: Principal Findings, Applications, and Research Frontiers* (pp. 147–186). Cambridge, MA: MIT Press.

Garrick, D., Schlager, E., & Villamayor-Tomas, S. (2016). Governing an International Transboundary River: Opportunism, Safeguards, and Drought Adaptation in the Rio Grande. *Publius, 46*(2), 170–198.

Gerlak, A. K., Lubell, M., & Heikkila, T. (2013). The Promise and Performance of Collaborative Governance. In M. E. Kraft & S. Kamieniecki (Eds.), *The Oxford Handbook of US Environmental Policy* (pp. 413–434). New York, NY: Oxford University Press.

Gleick, P. H. (1998). The Human Right to Water. *Water Policy, 1*(1998), 487–503.

Guerrero, A. M., Bodin, Ö., McAllister, R. R. J., & Wilson, K. A. (2015). Achieving Social-Ecological Fit Through Bottom-Up Collaborative Governance: An Empirical Investigation. *Ecology and Society, 20*(4), 41.

Halbe, J., Pahl-Wostl, C., Sendzimir, J., & Adamowski, J. (2013). Towards Adaptive and Integrated Management Paradigms to Meet the Challenges of Water Governance. *Water Science and Technology, 67*(11), 2651–2660.

Henry, A. D. (2011). Ideology, Power, and the Structure of Policy Networks. *Policy Studies Journal, 39*(3), 361–383.

Hering, J. G., & Ingold, K. (2012). Water Resources Management: What Should Be Integrated? *Science, 336*(6086), 1234–1235.

Herzog, L. M., & Ingold, K. (2019). Threat to Common-Pool Resources and the Importance of Forums: On the Emergence of Cooperation in CPR Problem Settings. *Policy Studies Journal*, online.

Hooghe, L., & Marks, G. (2001). Types of Multi-Level Governance. *European Integration Online Papers, 5*(11), 1–31.

Huxham, C., Vangen, S., Huxham, C., & Eden, C. (2000). The Challenge of Collaborative Governance. *Public Management an International Journal of Research and Theory, 2*(3), 337–358.

Iacobucci, D. (1996). *Networks in Marketing*. London: Sage.

Ingold, K. (2011). Network Structures Within Policy Processes: Coalitions, Power, and Brokerage in Swiss Climate Policy. *Policy Studies Journal, 39*(3), 435–459.

Ingold, K. (2014). How Involved Are They Really? A Comparative Network Analysis of the Institutional Drivers of Local Actor Inclusion. *Land Use Policy, 39*, 376–387.

Ingold, K., & Fischer, M. (2014). Drivers of Collaboration to Mitigate Climate Change: An Illustration of Swiss Climate Policy Over 15 Years. *Global Environmental Change, 24*, 88–98.

Ingold, K., Fischer, M., de Boer, C., & Mollinga, P. P. (2016). Water Management Across Borders, Scales and Sectors: Recent Developments and Future Challenges in Water Policy Analysis. *Environmental Policy and Governance, 26*(4), 223–228.

Ingold, K., Moser, A., Metz, F., Herzog, L., Bader, H. P., Scheidegger, R., & Stamm, C. (2018). Misfit Between Physical Affectedness and Regulatory Embeddedness: The Case of Drinking Water Supply Along the Rhine River. *Global Environmental Change, 48*, 136–150.

Ingold, K., Driessen, P. P. J., Runhaar, H. A. C., & Widmer, A. (2018). On the Necessity of Connectivity: Linking Key Characteristics of Environmental Problems with Governance Modes. *Journal of Environmental Planning and Management*, online.

Jager, N. W. (2016). Transboundary Cooperation in European Water Governance – A Set-Theoretic Analysis of International River Basins. *Environmental Policy and Governance, 26*(4), 278–291.

Jønch-Clausen, T. (2004). Integrated Water Resources Management (IWRM) and Water Efficiency Plans by 2005: Why, What and How? Global Water Partnership, Stockholm. Retrieved from www.gwptoolbox.org/images/stories/gwplibrary/background/tec_10_english.pdf.

Jordan, A., & Lenschow, A. (2010). Environmental Policy Integration: A State of the Art Review. *Environmental Policy and Governance, 20*(3), 147–158.

Kallis, G., & Butler, D. (2001). The EU Water Framework Directive: Measures and Implications. *Water Policy, 3*, 125–142.

Koppenjan, J. F. M., & Klijn, E. H. (2004). *Managing Uncertainties in Networks: A Network Approach to Problem Solving and Decision Making*. London: Routledge.

Kossinets, G. (2006). Effects of Missing Data in Social Networks. *Social Networks, 28*(3), 247–268.

Lafferty, W. M., & Hovden, E. (2003). Environmental Policy Integration: Towards an Analytical Framework. *Environmental Politics, 12*(3), 1–22.

Leifeld, P., & Schneider, V. (2012). Information Exchange in Policy Networks. *American Journal of Political Science, 56*(3), 731–744.

Leifeld, P., Cranmer, S. J., & Desmarais, B. A. (2018). Temporal Exponential Random Graph Models with btergm: Estimation and Bootstrap Confidence Intervals. *Journal of Statistical Software, 83*(6).

Lubell, M. (2003). Collaborative Institutions, Belief-Systems, and Perceived Policy Effectiveness. *Political Research Quarterly, 56*(3), 309–323.

Lubell, M. (2013). Governing Institutional Complexity: The Ecology of Games Framework. *The Policy Studies Journal, 41*(3), 537–559.

Lubell, M., & Edelenbos, J. (2013). Integrated Water Resources Management: A Comparative Laboratory for Water Governance. *International Journal of Water Governance, 1*(2013), 177–96. https://doi.org/10.7564/13-IJWG14

Lubell, M., Henry, A. D., & McCoy, M. (2010). Collaborative Institutions in an Ecology of Games. *American Journal of Political Science, 54*(2), 287–300.

Lubell, M., Scholz, J., Berardo, R., & Robins, G. (2012). Testing Policy Theory with Statistical Models of Networks. *Policy Studies Journal, 40*(3), 351–374.

Maag, S., & Fischer, M. (2018). Why Government, Interest Groups, and Research Coordinate: The Different Purposes of Forums. *Society and Natural Resources, 31*(11), 1248–1265.

Margerum, R. D. (2008). A Typology of Collaboration Efforts in Environmental Management. *Environmental Management, 41*(4), 487–500.

McGinnis, M. D., & Ostrom, E. (2014). Social-Ecological System Framework: Initial Changes and Continuing Challenges. *Ecology and Society, 19*(2), 30.

McPherson, M., Smith-Lovin, L., & Cook, J. M. (2001). Birds of a Feather: Homophily in Social Networks. *Annual Review of Sociology, 27*(1), 415–444.

Meadowcroft, J. (2009). What About the Politics? Sustainable Development, Transition Management, and Long Term Energy Transitions. *Policy Sciences, 42*(4), 323–340.

Metz, F., & Fischer, M. (2016). Policy Diffusion in the Context of International River Basin Management. *Environmental Policy and Governance, 26*(4), 257–277.

Newig, J. (2012). More Effective Natural Resource Management Through Participatory Governance? Taking Stock of the Conceptual and Empirical Literature – And Moving Forward. In K. Hogl, E. Kvarda, R. Nordbeck, & M. Pregernig (Eds.), *Environmental Governance. The Challenge of Legitimacy and Effectiveness* (pp. 46–68). Cheltenham: Edward Elgar.

Nilsson, M., Griggs, D., & Visbeck, M. (2016). Policy: Map the Interactions Between Sustainable Development Goals. *Nature, 534*(7607), 320–322.

Opsahl, T., Agneessens, F., & Skvoretz, J. (2010). Node Centrality in Weighted Networks: Generalizing Degree and Shortest Paths. *Social Networks, 32*(3), 245–251.

Pahl-Wostl, C., Holtz, G., Kastens, B., & Knieper, C. (2010). Analyzing Complex Water Governance Regimes: The Management and Transition Framework. *Environmental Science and Policy, 13*(7), 571–581.

Pärli, R., & Fischer, M. (2020). Implementing the Agenda 2030 – What Is the Role of Forums? *International Journal of Sustainable Development and World Ecology.* Online first.

Prell, C., Hubacek, K., & Reed, M. (2009). Stakeholder Analysis and Social Network Analysis in Natural Resource Management. *Society and Natural Resources, 22*(6), 501–518.

Raadgever, G.T.T., Dieperink, C., Driessen, P.P.J, Smit, A.A.H., van Rijswick, H.F.M.W. (2011). Uncertainty Management Strategies: Lessons from the Regional Implementation of the Water Framework Directive in the Netherlands. *Environmental Science and Policy* 14 (1): 64–75.

Robins, G., Snijders, T., Wang, P., Handcock, M., & Pattison, P. (2007). Recent Developments in Exponential Random Graph (p*) Models for Social Networks. *Social Networks, 29*(2), 192–215.

Rogers, P., & Hall, A. W. (2003). *Effective Water Governance* (Vol. 7). Stockholm: Global Water Partnership.

Sarewitz, D., & Pielke, R. A., Jr. (2007). The Neglected Heart of Science Policy: Reconciling Supply of and Demand for Science. *Environmental Science and Policy, 10*(1), 5–16.

Sayles, J. S., Mancilla Garcia, M., Hamilton, M., Alexander, S. M., Baggio, J. A., Fischer, A. P., Ingold, K., Meredith, G. R., & Pittman, J. (2019). Social-Ecological Network Analysis for Sustainability Sciences: A Systematic Review and Innovative Research Agenda for the Future. *Environmental Research Letters, 14*(9). https://doi.org/10.1088/1748-9326/ab2619.

Scholz, J. T., Berardo, R., & Kile, B. (2008). Do Networks Solve Collective Action Problems? Credibility, Search, and Collaboration. *The Journal of Politics, 70*(2), 393–406.

Sivapalan, M., Savenije, H. H. G., & Blöschl, G. (2012). Socio-Hydrology: A New Science of People and Water. *Hydrological Processes, 26*(8), 1270–1276.

Smith, J. A., & Moody, J. (2013). Structural Effects of Network Sampling Coverage I: Nodes Missing at Random. *Social Networks, 35*(4), 652–668.

Snijders, T. A., Van de Bunt, G. G., & Steglich, C. E. (2010). Introduction to Stochastic Actor-Based Models for Network Dynamics. *Social Networks, 32*(1), 44–60.

Stadtfeld, C., Hollway, J., & Block, P. (2017). Dynamic Network Actor Models: Investigating Coordination Ties Through Time. *Sociological Methodology, 47*(1), 1–40.

Tashakkori, A., & Teddlie, C. (2010). *Sage Handbook of Mixed Methods in Social and Behavioral Research*. London: Sage.

Tosun, J., & Lang, A. (2017). Policy Integration: Mapping the Different Concepts. *Policy Studies, 38*.

Treml, E. A., Fidelman, P. I. J., Kininmonth, S., Ekstrom, J. A., & Bodin, Ö. (2015). Analyzing the (Mis)Fit Between the Institutional and Ecological Networks of the Indo-West Pacific. *Global Environmental Change, 31*, 263–271.

Ulibarri, N., & Scott, T. A. (2017). Linking Network Structure to Collaborative Governance. *Journal of Public Administration Research and Theory, 27*(1), 163–181.

United Nations. (2015). Resolution Adopted by the General Assembly on 25 September 2015. Retrieved May 27, 2018, from http://www.un.org/ga/search/view_doc.asp?symbol=A/RES/70/1andLang=E.

Valente, T., Coronges, K., Lakon, C., & Costenbader, E. (2008). How Correlated Are Network Centrality Measures? *Connections, 28*(1), 16.

van Enst, W., Driessen, P. P. J., & Runhaar, H. A. C. (2014). What Works Where, When and How? Towards a Research Agenda. *Journal of Environmental Assessment Policy and Management, 16*(1), 1–25.

Varone, F., Narath, S., Aubin, D., & Gerber, J. D. (2013). Functional Regulatory Spaces. *Policy Sciences, 46*(4), 311–333.

Victor, J. N., Montgomery, A. H., & Lubell, M. (2017). *The Oxford Handbook of Political Networks*. Oxford: Oxford University Press.

Vignola, R., McDaniels, T. L., & Scholz, R. W. (2013). Governance Structures for Ecosystem-Based Adaptation: Using Policy-Network Analysis to Identify Key Organizations for Bridging Information Across Scales and Policy Areas. *Environmental Science and Policy, 31*, 71–84.

Vogel, J. M. (2004). Tunnel Vision: The Regulation of Endocrine Disruptors. *Policy Sciences, 37*(3), 277–303.

Weible, C. M., & Sabatier, P. A. (2005). Comparing Policy Networks: Marine Protected Areas in California. *Policy Studies Journal, 33*(2), 181–201.

Wasserman, S., & Faust, K. (1994). *Social Network Analysis Methods and Applications* (Vol. 8). Cambridge: Cambridge University Press.

Weingartner, R., Barben, M., & Spreafico, M. (2003). Floods in Mountain Areas – An Overview Based on Examples from Switzerland. *Journal of Hydrology, 282*, 10–23.

Weitz, N., Carlsen, H., Nilsson, M., & Skånberg, K. (2018). Towards Systemic and Contextual Priority Setting for Implementing the 2030 Agenda. *Sustainability Science, 13*, 531–548.

Widmer, A., Herzog, L., Moser, A., & Ingold, K. (2019). Multilevel Water Quality Management in the International Rhine Catchment Area: How to Establish Social-Ecological Fit Through Collaborative Governance. *Ecology and Society, 24*(3), 27.

Wolf, A. T., Yoffe, S., & Giordano, M. (2003). International Waters – Identifying Basins at Risk. *Water Policy, 5*, 29–60.

3

Network Segregation and Water Governance: The Case of the Spiny Lobster Initiative

Matthew Robbins and Mark Lubell

Introduction

Segregation into different subgroups is one of the most commonly observed and analyzed phenomena in environmental policy network studies (Freeman 1978; Henry 2017). For example, Leifeld and Schneider (2012) found that interest groups working on German toxic chemical issues were more likely to exchange strategic and technical information with other interest groups than with other types of organizations in the network. Gerber et al. (2013) found municipal governments collaborating on regional planning efforts tended to collaborate with other cities having similar levels of urbanization, similar values on socioeconomic variables, and with populations holding similar political beliefs. In the Central Kalimantan forestry policy arena, Gallemore et al. (2014) found that a shared scope of operations and similarity in opinions was associated with an increased likelihood of collaboration. And Alexander et al.

M. Robbins (✉) • M. Lubell
University of California, Davis, CA, USA
e-mail: mjrobbins@ucdavis.edu; mnlubell@ucdavis.edu

© The Author(s) 2020
M. Fischer, K. Ingold (eds.), *Networks in Water Governance*, Palgrave Studies in Water Governance: Policy and Practice, https://doi.org/10.1007/978-3-030-46769-2_3

(2018) found that networks of fishers in Jamaica's small-scale fishery were segregated by the type of fishing gear they employed. These examples, along with others cited throughout this chapter, demonstrate the importance of network segregation for environmental policy and governance.

In this chapter, we discuss five important dimensions of network segregation that have been emphasized in the literature: geography, policy beliefs, social group identity, sectoral affiliations, and trust/reciprocity. Each of these mechanisms of network segregation has different micro-level drivers that may entail subgroup formation. While these drivers may be correlated with each other depending on the context, it is often the case that the drivers produce cross-cutting social pressures to identify with different subgroups. This is one reason why in policy networks, individual people or organizations might identify with multiple different subgroups with various degrees of overlap.

Network segregation can be interpreted from the social capital perspective. A long-standing theme within the broader network literature is the distinction between "closed" and "open" networks, which social capital theory calls "bonding" and "bridging" social capital (Burt 2005; Geys and Murdoch 2008). The processes producing network segregation are mainly involved with the development of bonding social capital. Network segregation is driven by shared interests or some type of common social identity and it results in the formation of subgroups (Henry et al. 2011; Jackson 2014). These shared interests and identities produce a tendency for homophily in network relationships (Mcpherson et al. 2001). Research on cooperation provides a vast amount of evidence that cooperation is facilitated by group identity, which is supported by reciprocity and transitive network structures (Burt 2005; Berardo and Scholz 2010). In other words, the same structural features of networks that produce subgroups are also hypothesized to support social capital and cooperation within groups.

Network segregation has important consequences for theories of collaborative water governance, which Emerson et al. (2012, p. 2) define as "the processes and structures of public policy decision making and management that engage people constructively across the boundaries of public agencies, levels of government, and/or the public, private and civic spheres in order to carry out a public purpose that could not otherwise be accomplished". On the one hand, collaborative governance seeks to build bonding social capital and cooperation among policy stakeholders with interests in some water resource. In contrast, collaborative governance is wary of subgroups that form into competing advocacy coalitions, which fans conflict (Sabatier and Jenkins-Smith 1993) and produces echo-chamber and group-think effects within subgroups (Jasny et al. 2015). Hence, collaborative governance also values bridging social capital that builds relationships between subgroups, in order to share information or expand the basis of cooperation (Bodin and Crona 2009; Laird-Benner and Ingram 2010). Understanding the dynamics of collaborative governance thus requires understanding how network segregation and the attendant balance of bridging and bonding social capital changes over time, and the implications for trust, conflict, and information sharing.

This chapter explores ideas of network segregation in the context of the Spiny Lobster Initiative (SLI) in Honduras, which adopted a collaborative governance approach to develop an integrated partnership for spiny lobster management. Spiny lobster fisheries in the Caribbean and Pacific are managed for both local consumption and export, and like many other global fisheries, feature a tight policy monopoly (Baumgartner and Jones 1991) among commercial fishing operations, regional fishing management authorities, and the export sector. These fisheries often rely on exploitative labor practices among local populations. The goal of programs like the SLI is to develop more collaborative and equitable fisheries governance arrangements by integrating sectors that were previously excluded from the policy monopoly, but which nevertheless had interests in fisheries management outcomes. In the case of the SLI, this meant integrating other types of local government agencies, environmental groups, indigenous groups, and international NGOs. From the network segregation perspective, the SLI sought to build bridging social capital between subgroups formed mainly on the basis of geographic and sectoral identities.

The remainder of this chapter takes on the following tasks. We first provide an overview of different theoretical drivers of network segregation and how they are generally measured and analyzed in social and policy network literature. This includes a discussion about the distinction between exogenous identification of subgroups versus endogenous approaches to community detection. We then zoom in on some hypotheses about geographic and sectoral drivers of segregation in the case of the SLI, as bridging these two sources of segregation was a main goal of the program. The analysis uses network data from organizations involved in the SLI to examine how patterns of network segregation have changed over time. The conclusion discusses the lessons learned from the SLI analysis about the difficulty of changing processes of network segregation.

Theory

In this section, we discuss the drivers of the five most common sources of network segregation examined in the environmental policy network literature: geography, policy beliefs, social group identity, sectoral affiliation, and trust/reciprocity. Table 3.1 summarizes our basic points about the theory and analysis of these five mechanisms. The drivers of network segregation are not mutually exclusive; in some instances they may reinforce each other, while in other cases they may lead to cross-cutting patterns of segregation. Each subsection provides more explanation of the

Table 3.1 Dimensions of network segregation

Dimension of segregation	Driving mechanism	Measurement
Geography	Common biophysical impacts or geographic identity	Geographic proximity or co-location
Policy beliefs and preferences	Shared policy beliefs, attitudes, and preferences	Surveys or computational approaches to measuring beliefs, attitudes, and preferences
Social group identification	Common cultural values and norms	Ethnicity, profession, or other social markers
Sectoral affiliations	Shared economic or political sectors	Explicit role in market or political processes
Trust and reciprocity	Reciprocal interactions between trustworthy actors	Attitudes of trust, and expectations of promise keeping; network motifs of reciprocity and transitivity

process, identifies examples from water governance literature (not meant to be an exhaustive review), and offers ideas about analysis and measurement.

Geography

Geographical segregation may be driven by shared experiences with a particular problem, or by geographic place attachment and co-location. Geographical proximity has been shown to be associated with collaboration (Mcpherson et al. 2001; Kossinets and Watts 2009; Lubell et al. 2014). Individuals in the same geography might share the same risks from climate change, the same common pool resource, or generate or receive a positive or negative externality. Because of the link between geographic location and some particular environmental problem, the individuals and organizations within that geography may develop shared policy preferences or interests. For example, they may participate in political or policy processes to represent a particular geographic location.

The shared problem dynamics may be accompanied by strong place attachment that is integral to the social and cultural identity of a community. For example, people in San Francisco have strong attachment to the "Bay Area", which helps them work together and also creates incentives to represent the Bay Area in California or national-level politics in the United States. Actors also sort along multiple levels of geographic scale, from local to global. For example, Lubell et al. (2014) find that two actors that work in the same county are more likely to participate in the same water management policy forum.

Geographic segregation is easily measured by physical distances between actors' locations or by grouping actors according to where they fall within administrative jurisdictional boundaries. Actors who are geographically closer, or share the same geographic place identity, are expected to have higher densities of network relationships, reciprocity, and triadic closure. In statistical models like exponential random graph models (ERGMs) or stochastic actor-oriented models (SAOMs), geographic segregation would be indicated by a positive coefficient on homophily terms for sharing the same geographic location.

Policy Beliefs and Preferences

The Advocacy Coalition Framework (Sabatier and Jenkins-Smith 1993) argues that policy actors have three-tier belief systems consisting of fundamental deep core beliefs, policy-core beliefs embodying general preferences within a policy domain, and secondary-beliefs about specific policy issues. The belief systems are hierarchical; the more fundamental beliefs are hard to change and constrain the information processing and cognitive processes that shape the development of secondary beliefs. From a policy process and network segmentation standpoint actors with similar belief system form "advocacy coalitions" that coordinate participation in multiple political venues to pursue common policy goals. Competing advocacy coalitions are a source of political conflict, and one goal of collaborative governance is to create relationships that span coalition boundaries in order to reduce distrust and increase empathy.

Weible and Sabatier (2005) demonstrate how policy coordination networks in California segregate into two coalitions based on pro- and anti-Marine Protected Area beliefs. Henry et al. (2010) demonstrate that transportation policy stakeholders with similar policy beliefs are more likely to collaborate. These advocacy coalitions are supported by network structures associated with bonding social capital and network segregation. In contrast, Calanni et al. (2014) find that coalition formation within collaborative aquaculture partnerships is not driven by policy preferences, but rather is more influenced by trust and actor resources. This suggests a potential interaction between institutional structure and the relative importance of policy beliefs in driving political behaviors like coalition formation.

Shared attitudes, policy beliefs, or preferences have mostly been measured with surveys, but are also increasingly being measured with natural language processing or other automated data science approaches applied to some type of archival data (e.g., meeting minutes, plans, social media, news articles). The typical approach is to measure the extent to which an individual actor expresses a set of beliefs or policy preferences, which could be measured as binary (e.g., support/oppose a particular policy) or ordered categorical variables (e.g., Likert-scales ranging from pro-environmental to anti-environmental). Based on the individual answers, it is easy to develop various measures of "belief similarity" for pairs of actors, which can then be correlated (or statistically modeled) with the presence/absence of a network relationship.

Social Group Identity

Social group identity is one of the longest-standing ideas in social science. Centola et al. (2007) identifies processes of homophily and social influence that interact to maintain social group identity and cohesion. There are two forms of "choice homophily", meaning that actors' behavior is driven by their preferences to interact with other similar actors. One type of choice homophily, known as "value homophily", is the phenomenon whereby actors feel justified in their own opinions when surrounded by others sharing those same opinions (Lazarsfeld and Merton 1954; Knoke 1990; McPherson et al. 2001). The second type of choice homophily is known as "status homophily" and refers to instances where patterns of interaction are driven by actors' preference for others with similar cultural backgrounds (Lazarsfeld and Merton 1954). A third factor is a form of induced homophily whereby people become more similar to others the more they interact act with them (Smith-Lovin and McPherson 1987; McPherson et al. 2001). This form of homophily is also known as "social influence". Homophily and social influence are processes that can give rise to network segregation.

This broad social science literature is complimented by cultural evolution theory from anthropology, which focuses on the importance of social learning strategies among group members. Ethnic and group markers play an important role in defining which individuals are eligible targets of social learning (Boyd and Richerson 1987), and also contribute to group conflict and multi-level selection (Henrich and Boyd 1998). Cultural evolution theory posits a "social tribal instincts" hypothesis, where individuals are more likely to cooperate with members of the same group, and less likely to cooperate with out-group members (Boyd and Richerson 2009). These basic social instincts also undergird advocacy coalitions, as well as many other processes of group identity and segregation in society.

Barnes et al. (2016) (see also Barnes-Mauthe et al. 2013) provide a recent empirical example of how ethnic identity influences network segregation. Based on extensive interviews with fishing vessel captains, they find that the Hawaiian long-line fishing network is segregated into three ethnic groups: Vietnamese Americans, European Americans, and Korean Americans. This network segregation limits the diffusion of more sustainable fishing practices, which they estimate could have "prevented the incidental catch of 46,000 sharks between 2008 and 2012" (p. 6467).

Social group identity is fairly simple to capture from a measurement perspective. Individuals can be assigned to various ethnicity categories, for example using US Census definitions or local cultural categories. There are also many other potential markers of group identity that are relevant for environmental policy: different professional affiliations or scientific disciplines, local versus outsiders, environmentalist versus industry, and others relevant to a particular context. Obviously, there are long-running and politically fraught fights about how these categories are defined, but for the proximate purposes of network analysis, the researcher will have to choose a clear categorization to support quantitative modeling. For group identifiers that are unique to a specific context, the analyst is required to develop enough local qualitative knowledge to effectively identify group markers and measure them with surveys or other types of quantitative instruments. For example, a survey could include a checklist of group categories that an individual can select, or the analyst could use expert judgment to sort actors into relevant social group categories.

Sectoral Affiliations

Economic and political systems are characterized by a variety of specialized sectors with defined roles for individual actors. Working within a familiar sector is easier to manage and may be less resource intensive, whereas working across boundaries increases opportunity costs and increases risk of failure. Economic sectors might be broad categories like "labor" and "management" (Barnes et al. 2016), but are often divided into more specialized components in the context of environmental policy or even within organizations (Cross et al. 2002). For example, researchers might examine different types of energy producers, farmers, or commercial versus recreational fisheries. These different economic sectors usually have different policy preferences depending on their particular role in systems of economic exchange and production.

Political systems are also characterized by specialized sectors and roles, which might be as broad as "governmental" versus "non-governmental" actors, or "governmental" versus "interest group" versus "research" (Maag and Fischer 2018). But again, there are more specialized sectors defined by institutional arrangements, jurisdictions, and policy interests. For example, government agencies have jurisdiction for different aspects of environmental issues, like fisheries, forests, energy, and land-use, which often creates interagency conflict and fragmentation at the ecosystem level. Regional government actors often have a different perspective than local government actors when dealing with regional problems (Gerber et al. 2013). Non-governmental actors like interest groups can also be divided up into sectors, for example environmental justice groups versus wilderness advocacy groups. These types of interest-based political sectors are often aligned with ideological and belief-system differences.

Lienert et al. (2013) provide an example of both horizontal and vertical segregation in the realm of water infrastructure planning. The authors found a dearth of horizontal cooperation between the water supply and wastewater sectors, as well as few ties between government actors at different decisional levels. Fischer et al. (2019) found that even in the context of water forums, institutions specifically designed to encourage cross-sectoral interactions, homophilic tendencies led forum composition to be strongly segregated by actor sector (private/public/scientific). Emerging research on the water-energy-food nexus suggests these three sectors, which are linked by various biophysical processes, have extremely "silo-ed" networks with limited cross-sectoral collaboration (Kurian et al. 2018).

Measuring sectoral functions usually requires specialized knowledge of the particular research context. The relevant economic sectors are often defined by the industrial organization of a particular economic activity, such as the value chains associated with agricultural production or natural resource development. For instance, the Honduran spiny lobster fishery includes boat captains, indigenous groups that provide labor, local sea food purchasers, and international exporters. The industrial organization of an economic activity is often linked to the political structures, where different government agencies may have direct authority and jurisdiction over different aspects of the economic activity, while other agencies may be concerned with social and economic outcomes that are indirectly associated with the activity. The same type of direct and indirect interests may also define boundaries among non-governmental actors, who operate at different levels of geographic scale (i.e., local to international).

Trust and Reciprocity

Trust and reciprocity are widely recognized as the core ingredients of bonding social capital and the foundation for cooperation (Axelrod 1984; Coleman 1988; Putnam 2001). Trust and reciprocity also cause networks to coalesce in subgroups, as actors choose to interact with others on the basis of reputation. Trust can be conferred by exogenous markers like social identity or geographic origin, but trust can also emerge from a history of interaction that is independent of these other mechanisms of network segregation. A key mechanism behind the evolution of cooperation is the clustering of reciprocal strategies—or "positive assortment"—whereby cooperative strategies are more likely to interact with each other (Axelrod 1984; Nowak 2006; Smaldino and Lubell 2011). Experiments have shown that actors with a reputation for trusting each other tend to self-select into cooperation games with other trustworthy actors, and banish less non-cooperative actors to peripheral groups (Ahn et al. 2009). These processes are complimented by transitivity, or "friend of a friend" processes, where one actor brokers a relationship between two other actors on the basis of reputation (Faust 2010).

Many research projects in water governance have examined trust and reciprocity, especially in the context of collaborative governance where often the goal is to develop trust across boundaries formed by other mechanisms of network segregation. For example, Calanni et al. (2014) found that in a marine aquaculture policy network composed of both governmental and non-governmental organizations, professional competence, an aspect of trust, was a strong predictor of selection for collaborative partnerships. In a set of ten estuary water policy arenas, Berardo and Scholz (2010) found that actors tended to partner with popular, boundary-spanning organizations under low-risk conditions, while trust was a more important determinant of collaboration in high-risk contexts. Ulibarri and Scott (2017) compared levels of collaboration for three federal hydropower relicensing processes, finding that the process with the highest level of collaboration also had the highest level of reciprocity and the lowest density of connections. Lubell et al. (2014) found that coordination in the San Francisco Bay water management policy arena was facilitated primarily by government actors and by geographic boundary-spanning collaborative institutions, all three of which were associated with high rates of network closure.

Trust and reciprocity are often measured on the basis of survey questions that ask individuals to evaluate the trustworthiness of other specific actors in the policy context, or more aggregated groups (e.g., how much do you trust other stakeholders involved in this partnership?). The analysis then might test how much trust predicts partner selection in networks, or the likelihood of accessing different types of information sources (Lubell 2007). Another approach is to use network analysis methods such as exponential random graph models to estimate the tendency of particular types of network motifs such as reciprocal ties or transitivity, which are theorized to be indicators of cooperation.

Empirical Case Study: The Spiny Lobster Initiative in Honduras

In this section, we provide an empirical example of the role of network segregation in the context of the Spiny Lobster Initiative in Honduras. We first provide a basic description of the SLI in order to establish the context. We then summarize some hypotheses about network segregation based on geography and sector, which were the two most important mechanisms of network segregation in the context of the SLI. The hypotheses translate the more general discussion of mechanisms above into the specific context of the SLI. Finally, we summarize the collection of several types of network data, along with descriptive statistics and statistical models of network structure.

The Spiny Lobster Initiative as Collaborative Governance

The spiny lobster (*Panulirus argus*) fishery in Honduras is worth nearly US $50 million in exports to the United States annually and provides direct employment to more than 4000 people from coastal communities. It is Honduras' most valuable wild caught fishery and the second most valuable spiny lobster fishery in Central America after Nicaragua. Yet, despite its economic importance, national management strategies and weak fisheries governance have made the fishery unsustainable (Christie 2014). In addition, SCUBA dive fishing, primarily conducted by indigenous Miskito men, is one of the principal ways that lobster is caught and threatens divers, lobster stocks, and biodiversity. Honduras is one of the poorest countries in Latin America and La Moskitia is its most impoverished region. As lobster stocks decrease, fishers dive deeper, longer, and more frequently increasing the incidence of injury and mortality (Christie 2014). Partly due to their cultural and geographic isolation, Miskito divers and community have historically lacked representation in the management of the Honduran spiny lobster fishery.

In 2009, the Global Fish Alliance, an international development partnership between the United States Agency for International Development (USAID), Darden Restaurants (Red Lobster), and the human development organization FHI360, began the Spiny Lobster Initiative (SLI). Through the SLI, the Global Fish Alliance sought to "enhance livelihoods, biodiversity and food security by promoting sustainable fisheries and responsible aquaculture … [and] centered on the application of a system-wide approach that balances economic, environmental, governmental, and social issues essential to enhancing livelihoods and biodiversity". The SLI represented an application of the System-wide Collaborative Action for Livelihoods and the Environment (SCALE) methodology, a system-level approach to social change. SCALE is designed to achieve broad changes in stakeholder beliefs, attitudes, and practices and to facilitate collaboration across sectors of stakeholders along the value chain. As with other approaches to collaborative governance (Ansell and Gash 2007; Emerson et al. 2012), SCALE aims to build networks across subgroup boundaries in order to resolve conflict and pursue mutually beneficial activities.

Hypotheses: Geographic and Sectoral Function Segregation

A particular focus of the SLI was overcoming geographic (Fig. 3.1) and sectoral drivers of network segregation in the spiny lobster fishery. The spiny lobster fleet exploits the banks of Rosalinda, Gorda, Thunder Knoll, Media Luna, and Lagarto Reef, in the eastern territorial waters of Honduras. The fishery first developed on the Bay Islands that lay off the north coast of Honduras in the Caribbean Sea, more or less directly offshore from La Ceiba (Christie 2014). Being located on islands,

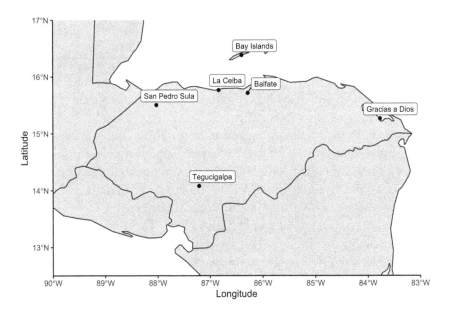

Fig. 3.1 Map of Honduras with locations associated with spiny lobster fishery organizations

organizations in the Bay Islands are clearly geographically isolated from the rest of the entities in the network that are based on the mainland. The Bay Islands are also distinct from the mainland in that they are populated mainly by English speakers, including white descendants of British settlers and the Garifuna people, black descendants of slaves. Lobster fishers on the Bay Islands have over time shifted away from SCUBA methods of catch to more lobster traps, in comparison to fishing communities on the mainland coast.

The mainland features two different geographical locations important for the spiny lobster fishery. First are the coastal areas like La Ceiba that have developed processing facilities and become the main home of the lobster fishing fleet; it was also the site of the SLI symposia. But a lot of the administrative resources are located in inland population centers like Tegucigalpa, which is the seat of the Honduran national government and home to the federal fisheries authorities. The network also contains a number of international organizations based outside Honduras, either involved with the spiny lobster fishing industry or as international development organizations. Within Honduras, these organizations are usually physically based in population centers like Tegucigalpa or La Ceiba, where more policy and economic interactions take place.

Perhaps the most important geographic location from the perspective of the SLI is Gracias a Dios, which is home to the indigenous Miskito lobster divers and artisanal fishers that provide the labor for the SCUBA fishing. Reducing the vulnerability and health risks to the Miskito divers was a central environmental justice goal for SLI. Gracias a Dios is an administrative department on the coast of Honduras that is isolated by lack of direct overland routes to La Ceiba and other cities to the West. Gracias a Dios contains the Rio Plata Bioreserve. Balfate is a Garifuna artisanal fishing community in the rural department of Colón on the coast east of La Ceiba and west of Gracias a Dios. The community remains unconnected from Honduras' highway system, separated from the nearest point by a ten-mile length of unpaved road.

Given the geographic structure of the spiny lobster fishery, and how the evolution of the industry varies across geography, we hypothesize a high level of geographic segregation.

Hypothesis 1 Geographic Segregation *As a result of proximity, actors are more likely to form ties with actors sharing their geographic location than with actors who are outside the same geography.*

The spiny lobster fishery and associated policy institutions can also be identified on the basis of different economic and political sectors. A main goal of the SLI was to bring together stakeholders from economic and political sectors that did not usually collaborate, but which made decisions that affected each other. Government entities tasked with managing the fishery (Government Direct), including the Fisheries Directorate and the Merchant Marines, participated as the holders of decision-making authority and enforcement responsibilities. The SLI aimed to support these organizations, which are widely recognized as lacking adequate resources to manage the fishery. Regional and local government entities without authority over spiny lobster fishing activities (Government Indirect) were included in recognition of the wide-ranging economic and social implications of fishery. These government organizations were engaged in broad efforts to manage existing knowledge, attitudes, and practices of not only fisheries but also food security, livelihoods, and biodiversity issues.

Industry groups along the supply chain representing interests at the resource extraction and the processing and exporting phases were included in the initiative. A main focus of the initiative was to implement a diving ban and provide an economic alternative for divers. As such, the SLI placed an emphasis on participation of the divers as well as artisanal fishers (Small Fishermen). The more economically and politically powerful lobster processors and exporters comprised the other industry sector (Processors and Exporters).

Environmental groups (Environment) operating at multiple scales from local to international represented the conservation interests. International aid agencies and NGOs participated as funders and organizers of the initiative (Donors and Cooperants). There were also a small number of additional organizations such as banks and media entities (Other) participating at a peripheral level.

Sectoral homophily refers to the tendency for actors operating within the same sectoral domain to share ties. For example, artisanal fishermen's associations from different regions might be more likely to communicate and collaborate with other such organizations and develop high levels of bonding social capital. Working within a familiar sector is easier to manage and may be less resource intensive, whereas working across boundaries increases opportunity costs and increases risk of failure. Stakeholders who are similar to one another are better able to communicate tacit, complex information, as there tends to be higher mutual understanding between such actors (Prell et al. 2009).

Hypothesis 2 Sectoral Segmentation *As a result of increased access and information, actors within the same sector are more likely to form ties with each other in comparison to actors from different sectors.*

A primary goal of the SLI was to create new relationships that spanned these existing sectoral and geographic boundaries. The geographic and sectoral boundaries were historically a source of conflict, but the economic and political processes operating in the spiny lobster fishery also created interdependences between these actors. At the very least, the SLI sought to open new lines of communication across subgroup boundaries with the possibility of joint activities or adjustment of individual decisions to better account for social costs and benefits. The political economic definition of cooperation is when individual actions account for interdependence. If the SLI was successful this goal, there would be evidence that geographic or sectoral segmentation would weaken over time.

Hypothesis 3 Collaborative Governance *As a result of increased cross-boundary communication, the magnitude of geographic and sectoral segregation will decrease over time.*

Methods and Analysis

This section first explains how we collected network data for organizational relationships with the SLI, and also identified sectoral and geographic attributes for each organization. We then present the results of descriptive network analysis to visualize the patterns of geographic and sectoral segmentation, and whether there is evidence that these processes may have changed over time in the SLI. Lastly, we conduct a more rigorous statistical model of network dynamics to see if geographic and sectoral segmentation is operating on the network while controlling for other important network processes.

Survey and Network Data Collection

A basic social network survey measuring the existence and strength of social relationships between organizations was administered during three waves of data collection. Two types of social relationships were assessed: frequency of communication and familiarity. Communication was measured on a 5-point scale ranging from "0 = We don't have contact or communicate with this organization" to "4 = Very frequently—weekly or more". Familiarity was measured on a 6-point scale ranging from "0 = We have not heard of this organization" to "5 = We have a contract or memorandum of understanding". Respondents answered each question for every target organization listed on the survey. The SLI and its core partners generated the initial list of organizations through the context mapping phase of the SCALE process. The list evolved over the course of the project as organizations disbanded or were identified as being active participants in the fishery.

Data was collected by administering the survey at three points in time. Wave 1 of the data set was collected at a "Whole System in the Room" stakeholder meeting in 2009 and subsequent lobster fishery technical symposium. Wave 2 was collected in 2011 by a consultant targeting organizations that had filled out a survey in Wave 1. Wave 3 was collected in 2013 at a final technical symposium to which all previous respondents had been invited. At each time point, paper-and-pen surveys were administered to participants who completed the survey on the spot. The survey included a sector of operation item, and SLI staff provided geographic location information.

This data collection process generated a set of directed, valued network ties for each Wave. The data was binarized by taking the valued data and mapping responses for each of the two network questions (relationship and familiarity) of 0, 1, or 2 to 0, and responses of 3 or higher to 1. If 1 was obtained on either question for a given tie, that tie was assigned a value of 1 indicating the existence of a connection between the two organizations. The network (N = 90) includes all of the organizations that at each time point either (a) filled out the survey and/or (b) were nominated by (i.e., received a connection from) one of the organizations that took a survey. In other words, an organization was included in the network if it either sent or received a tie at all three time points. For the purposes of this analysis, organizations that were not present at all three time points are excluded. Table 3.2 reports the percentage of organizations by sector and geography.

Network Visualization and Community Detection

Figure 3.2 visualizes the three waves of the network based on the endogenous communities identified by the Walktrap community detection algorithm. Walktrap community detection uses random walks to define the distance between vertices, and then uses a clustering algorithm to divide the network into subgroups (Pons and Latapy 2005). The nodes in the figure represent different organizations in the SLI, connected by information sharing links, and colored by cluster membership. The clusters are also visually defined by the colored polygons. Like other clustering algorithms, Walktrap sometimes splits off very small or single-member communities containing outliers, which makes it more informative to interpret the main larger communities. To interpret the results from the network segregation perspective, Figs. 3.3 and 3.4 report the composition of the communities in terms of economic sector and geographic location.

Wave 1 contains seven communities, although only four are large enough to provide meaningful interpretation. Community 1 (N = 5) consists of large US NGOs or Honduran federal government entities. Community 2 (N = 38) contains the rest of the international organizations, most of the remaining organizations from the Donors and Cooperants and the Environment sectors. The Global Fish Alliance (GFISH) and Darden Restaurants, the main funding and organizational

Table 3.2 Spiny Lobster Initiative network: organizations by sector and geography

SLI organizations by sector

	Donors and cooperants	Environment	Government indirect	Government direct	Processors and exporters	Small fishermen	Others
Number	12	13	27	6	6	16	10
Proportion of total (N = 90)	0.13	0.14	0.30	0.07	0.18	0.11	0.07

SLI organizations by geography

	Balfate	Bay Islands	Foreign	Gracias a Dios	La Ceiba	San Pedro Sula	Tegucigalpa
Number	3	13	7	15	29	2	21
Proportion of total (N = 90)	0.03	0.14	0.08	0.17	0.32	0.02	0.23

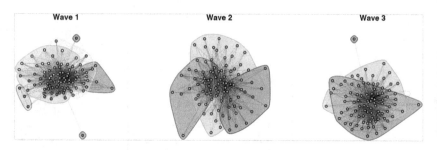

Fig. 3.2 Network community structure

partners in the SLI, are in this group. Community 3 ($N = 28$) includes mostly the fishing industry and directly associated government organizations such as the General Directorate for Fisheries and Aquaculture of Honduras (DIGEPESCA). Community 4 ($N = 14$) contains both divers associations and a plurality of the organizations are from the Gracias a Dios department. Geographically, there is a separation between the Bay Island and Gracias a Dios actors, and the actors in the capital of Tegucigalpa and the port of La Ceiba are more highly represented in the community containing Bay Island actors.

Wave 2 contains three meaningful communities; note that the order of the communities is not the same across the three time periods. Community 1 ($N = 22$) is the community with the most Small Fishermen sector organizations (including the divers associations), and excludes Government Direct and Processors/Exporters. Community 2 ($N = 53$) is the largest community containing all of the Government Direct and most of the Government Indirect and Donors and Cooperants sector organizations. Notable organizations include GFISH and Darden, as well as the Fisheries Directorate. Community 3 ($N = 15$) is the smallest community; it contains the majority of the Processors and Exporters and few other organizations. The geographic split between the Bay Island and Gracias a Dios actors is still evident, along with the association between the Bay Islands, the capital, and port.

Wave 3 contains two main communities, and continues the pattern of finding fewer clusters because there is less of a demarcation between groups. Community 1 ($N = 49$) contains all of the Processors and Exporters, most of the Donors and Cooperants including almost all of

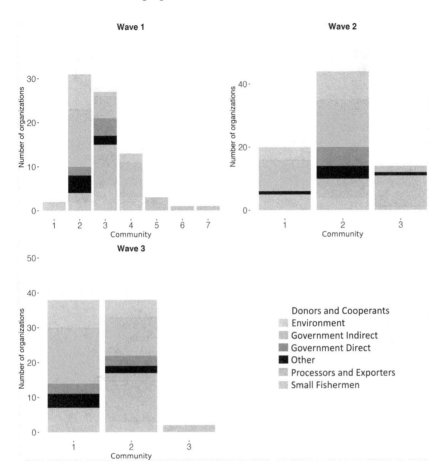

Fig. 3.3 Sectoral community detection results

the foreign organizations and all of the organizations located in the Bay Islands. Community 2 (*N* = 39) excludes all of the Processors/Exporters and has few Donors and Cooperants; it has a higher concentration of the small fishing organizations including the two divers associations. While the Bay Island and Gracias a Dios geographic segmentation is still evident, the capital and port actors are more evenly distributed across the two main communities.

Overall there is evidence of sectoral splitting between commercial fishing versus small-scale fishing, with Government Direct unusually

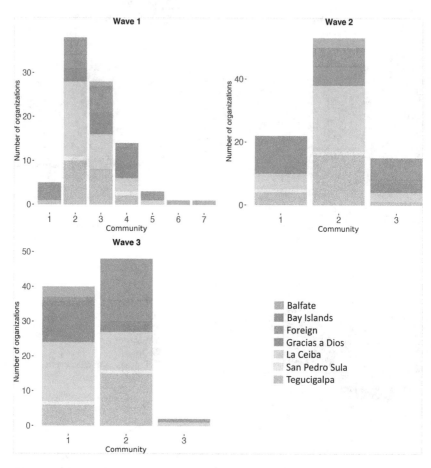

Fig. 3.4 Geographical community detection results

associated with fishing industry. Government Indirect plays an interesting role. These government organizations had limited previous interaction with groups in the fisheries policy subsystem and thus had an opportunity to form relationships with many new partners. Likely due to the administrative capacity of these organizations, they were able to capitalize on the opportunity afforded by the initiative. The increased integration of this sector into the lobster fishery network supports the initiative's goal of more broadly engaging government actors to affect broad societal change in attitudes and knowledge related to fishery.

In order to assess the degree to which network segregation by sector or by geographical location overlapped with network modularity revealed by community detection, we conducted chi-square tests of independence. To meet the minimum data requirements for a chi-square test, we grouped similar sectors to form three larger categories as follows: Government Direct and Government Indirect became a larger "Government" sector; Donors and Cooperants and Environmental became "NGOs"; and Processors and Exporters and Small Fishermen became "Industry". Similarly, for the geography variable we consolidated organizations into two large categories: a group of geographically-isolated organizations (Balfate, Gracias a Dios and Bay Islands) and a group consisting of well-connected mainland- and foreign-based organizations (Tegucigalpa, La Ceiba, San Pedro Sula, and Foreign). We also removed the identified structural communities containing three or fewer organizations.

The results of the chi-square tests between the sector attribute and identified Walktrap community membership are as follows: Wave 1, $\chi^2(4, N = 85) = 18.06$; $p < 0.01$; Wave 2 $\chi^2 (4, N = 90) = 22.49$, $p < 0.01$; and for Wave 3, $\chi^2 (2, N = 90) = 7.20$, $p < 0.05$. The chi-square tests indicate that sector and community membership was non-independent in all three time periods, but that after increasing from W1 to W2, the degree of non-independence declined strongly to W3. The results of the chi-square test between geographical location of operation and community membership are as follows: Wave 1, $\chi^2(2, N = 80) = 8.46$; $p < 0.05$; for Wave 2 $\chi^2(2, N = 90) = 17.99$, $p < 0.01$; and for Wave 3, $\chi^2 (1, N = 89) = 0.21$, $p = 0.65$. Similar to the results for the sector tests, by Wave 3 there is much weaker evidence of geographic segregation associated with the emergence of two subgroups instead of three.

Another way of measuring homophily in a social network is through the use of the External-Internal (E-I) index (Krackhardt 1998). The E-I index gives a measurement of the proportion of within-group and between group ties for a given group in a network. Specifically, the index value is the number of ties external to the group minus the number of ties that are internal to the group divided by the total number of ties involving actors in the group. The index ranges in value from −1 (all within

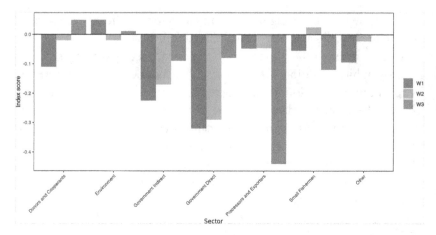

Fig. 3.5 External-internal index scores by sector over time

group ties) to +1 (all between group ties). Missing columns in Fig. 3.5 indicate that there is an E-I index of zero.

The E-I index results displayed in Fig. 3.5 reveal some clear trends for several of the sectors. Government Indirect had the most negative E-I value at the first time point, indicating the greatest difference between ties linking organizations within that sector vs. ties linking Government Indirect organizations to organizations in other sectors. The Government Indirect sector's E-I index score increased at each time point, starting at −0.22 and ending at −0.09. The only other sector whose E-I index increased over time is the Donors and Cooperants, which went from a negative E-I value to a positive value (−0.11 to 0.05). The Processors and Exporters E-I index score changed drastically, dropping off at Wave 3 to −0.44, the lowest proportion of outgoing ties from any sector at any time point. The Small Fishermen sector also exhibits a decreasing E-I index value over tie (−0.06 to −0.12).

The descriptive data provides at least some circumstantial evidence that sectoral and geographic processes of segregation were becoming weaker over time. Overall, the network became less modular, with fewer distinct communities at each time point. At Wave 3, the entire Processors and Exporters sector along with the key federal fisheries regulation agency appear in a community with the foreign NGOs and environmental groups that are funding and coordinating the program. There also appears to be some geographical segregation at the final time point, with very few (3 out of 15) organizations from Gracias a Dios appearing in Community 1 and none of the 13 Bay Islands organizations appearing in Community 2. The divers associations do not appear to be well-integrated into the network as they don't appear in the same community as the organizations with the most political and financial power. Drilling down to specific sectors, the Donors and Cooperants and the Government Indirect organizations were able to broaden their networks across sectoral lines over the course of the initiative. Meanwhile, both the Processors and Exporters and Small Fishermen sectors became more insular, with the number of within-sector ties increasing relative to cross-sector ties.

Stochastic Actor Oriented Models

We apply stochastic actor-oriented models (SAOMs) to provide a more rigorous statistical analysis of geographic and sectoral segregation. The behavior of actors results in continually evolving network configurations, and SAOMs for network dynamics give evidence about the evolution of social networks over time (Van De Bunt et al. 1999). They model network evolution based on the individual, rational choices made by actors over time as a Markov process (Snijders et al. 2010). Strong tendencies in tie formation, as formulated in our hypotheses, are represented by significant effects in the model, that is, prevalence of motifs in the network that cannot be due to random processes.

These models were estimated with the RSiena package in R (Ripley and Boitmantis 2010; Ripley et al. 2017). Results are deemed acceptable if the maximum convergence ratio is less than 0.25 and the convergence *t*-value for each parameter is smaller than 0.1, conditions which were met for all three models presented in Table 3.2. In addition, the Jaccard index, a measure of the amount of change between waves, exceeds 0.3 for both periods, yielding an acceptable model (Snijders et al. 2010).

An initial model (Model 1) includes endogenous terms to control for several network processes commonly hypothesized to affect network structure and indicate cooperation. *Density* establishes the baseline rate of tie formation. *Reciprocity* (a tie from A→B and from B→A) supports cooperation among actors (Berardo and Scholz 2010). *Transitive triplets* occur between three actors A, B, and C when there is a tie from A→B, from A→C, and from B→C; that is, the tendency for an actor to form a tie with another actor that shares a tie with a common third party. *Three cycles* occur when there is a cycle from A→B, B→C, and C→A (this link is reversed in comparison to transitive triplets). Transitive triplets and three-cycles are both related to group formation and bonding social capital (Carpenter et al. 2004; Block 2015). In-degree popularity reflects tendencies for actors with high numbers of incoming ties to attract extra ties "because" of their high current number of ties, a sort of "the-rich-get-richer" effect.

Models 2 and 3 then add geographic and sectoral homophily effects to test our two main hypotheses about network segregation. The sectoral homophily parameter checks for a match between the sector identity of the sending and receiving nodes. The geographic homophily effect checks for a match between the geographic location covariate attribute of the ego, or source of a tie, with the location attribute of the alters, or potential tie recipients. The geographic location attribute is a time-invariant covariate; each organization's location is the same at each wave of data collection. It is important to note that these models are not testing whether these processes of network segregation are becoming stronger or weaker over time; there is no interaction effect with time. Rather, the parameters capture of the average tendency of geographic and sectoral homophily to influence patterns of network change over time—the formation and dissolution of ties.

Model 3 also includes terms for outgoing (sociality) and incoming (popularity) ties for each sector. This helps determine whether geographic and sectoral homophily contributes to network segregation controlling for actor-level attributes that might attract a higher rate of network ties and be correlated with geography or sector.

Model Results

Table 3.3 reports the result of the SAOM analysis for all three models. The results strongly support our first and second hypotheses. Actors within the same sectors are more likely to form ties with each other than with actors from different sectors. Similarly, actors were more likely to form ties with other actors with a common geographic base of operations. The magnitude of the homophily parameters indicates they are among the strongest processes occurring on the SLI network.

There are also several interesting results from other network processes associated with the evolution of cooperation in the SLI. Reciprocity and transitivity are both positive forces in the network and indicate network closure and bonding social capital. At the same time, there is a negative coefficient on three-cycles, which some researchers suggest provide a non-hierarchical version of closure. The combination of a positive parameter for transitivity and a negative one for three-cycles indicate a strong hierarchical tendency in the network (Snijders et al. 2010). This apparent trend toward a more hierarchical structuring of relationships in the network runs counter to SLI's goals of fostering a more participatory governance arrangement in which communication and relationships are more horizontally distributed (Andino 2009). However, it is consistent with cooperation in the network requiring some type of centralized leadership (Provan and Kenis 2008). Furthermore, reciprocity and transitivity contribute to the modularity of the network and the emergence of endogenous subgroups, independent of exogenous attributes like sector and geography.

The potential importance of centralized leadership is supported by the results that government actors, donor and cooperant actors, as well as popular actors were preferentially targeted for the formation of new ties. Centralized organizational networks with a small number of nodes that occupy highly central positions may facilitate coordination as the central actors can distribute information efficiently to the rest of the network (Hagen et al. 1997; Turk 1977).

Table 3.3 SAOM model results

	Model 1	Model 2	Model 3
Network structure effects			
Density	−1.42**	−1.60**	−1.53**
	(0.09)	(0.08)	(0.11)
Reciprocity	0.65*	0.59**	0.5
	(0.30)	(0.22)	(0.38)
Transitive triplets	0.06**	0.06**	0.06**
	(0.01)	(0.00)	(0.00)
3-cycles	−0.07**	−0.06**	−0.05
	(0.02)	(0.02)	(0.04)
Indegree popularity	0.01**	0.01**	0.01
	(0.00)	(0.00)	(0.01)
Homophily			
Sector		0.29**	0.2**
		(0.06)	(0.07)
Geography		0.32**	0.38**
		(0.06)	(0.05)
Actor type—Outgoing			
Donors and cooperants			0.28*
			(0.13)
Environment			0.04
			(0.12)
Government indirect			0.15
			(0.12)
Government direct			−0.22
			(0.13)
Processors and exporters			−0.05
			(0.13)
Small fishermen			0.14
			(0.15)
Actor type—Incoming			
Donors and cooperants			0.01
			(0.12)
Environment			−0.08
			(0.11)
Government indirect			0.01
			(0.09)
Government direct			−0.32
			(0.13)
Processors and exporters			−0.15
			(0.10)
Small fishermen			0.00
			(0.10)

Levels of statistical significance: (*) 0.05, (**) 0.01

Conclusions

The overarching goal of this chapter was to catalog important mechanisms of network segregation for environmental governance, and then illustrate these mechanisms in operation for the Spiny Lobster Initiative. Perhaps the key take-home point from this discussion is how these five mechanisms of network segregation—geography, policy beliefs, social identity, sectoral affiliation, and trust/reciprocity—are extremely strong social processes that emerge from long histories of social and economic interactions. Geography is rooted in where people live and grow up; policy beliefs are shaped by fundamental ideological viewpoints; social identity is a core aspect of personality and social psychology; sectoral membership reflects enduring institutional structures and economic opportunities; trust and reciprocity depend on reputations built through long periods of interaction and are passed between generations. These processes reflect the distribution of political and economic power, and are often mutually reinforcing. While bonding social capital makes cooperation within subgroups more likely, the boundaries separating subgroups can become sources of conflict, and building cross-boundary relationships is difficult. Such processes are important for all types of environmental governance, not just water.

The SLI provides a good illustration of the enduring nature of network segmentation processes. Geographic and sectoral homophily were present throughout the entire SLI process. The geographic segmentation is driven by the physical distances between the islands, mainland, and the remote region that is home to the Moskito indigenous population. Geography played a role in shaping the economic development of the fishery, even including gear preferences. The fishery also featured a "policy monopoly" driven by large boats supplying an export market, and sympathetic government agencies with direct authority for fisheries management. Governmental and non-governmental stakeholders with indirect connections to the commercial fishing stakeholders, along with the disadvantaged indigenous communities, had more difficulty gaining access to the network.

There is only limited evidence that the SLI partnership was able to overcome these network segmentation processes, mostly by increasing the integration of government agencies that were indirectly involved with the fishery and linking to a broader set of donors and international organizations. These government organizations had limited previous interaction with groups in the fisheries policy subsystem and thus had an opportunity to form relationships with many new partners. Likely due to the administrative capacity of these organizations, they were able to capitalize on the opportunity afforded by the initiative. The increased integration of this sector into the lobster fishery network supports the initiative's goal of more broadly engaging government actors to affect broad societal change in attitudes and knowledge related to the fishery. However, the only other sector that became better integrated into the network, were the organizations associated with funding and facilitating the initiative. The presence of these organizations is limited to the duration of the initiative so the value of these relationships is temporary and underlines the "exit problem" commonly associated with international development projects. At the same time, the commercial fisheries stakeholders appeared to become even more insular, which may be linked to their political strategy to develop new policy forums that support their preferences.

When any particular network segmentation process is operating in a particular context, it is hard to imagine any collaborative governance process that would have a major effect in the short-term. Collaborative governance is usually a short-term policy intervention in the face of longer-term social processes, and any particular collaborative governance initiative like the SLI may compete with other programs in the same context (Lubell and Niles 2019; Rudnick et al. 2019). Perhaps longer-term change could be sparked by forming some new social relationships and lines of communication (Lubell and Lippert 2011), but longer-term cooperation may require reshaping of group identities. For example, instead of groups defined by local geography, a group identity at the regional level may facilitate more widespread cooperation. The appearance of reciprocity and transitivity as significant network processes may provide a reservoir of social capital that could be extended to new group identities.

Comparative analysis is an important agenda item for further research on network segmentation, and indeed all environmental and water governance processes. The nature and strength of network segmentation processes will vary over space and time, in some cases being weaker and others stronger. Sometimes segmentation processes will overlap and be mutually reinforcing, perhaps in concert with other network processes that drive modularity and subgroup formation. There may be contexts in which an institutional arrangement such as collaborative governance or some other policy tool may be more or less effective at changing processes of network segmentation in order to build cross-boundary cooperation. Only comparative research can provide definitive answers to these types of questions.

References

Ahn, T. K., Esarey, J., & Scholz, J. T. (2009). Reputation and Cooperation in Voluntary Exchanges: Comparing Local and Central Institutions. *The Journal of Politics, 71*(2), 398–413.

Alexander, S. M., Bodin, Ö., & Barnes, M. L. (2018). Untangling the Drivers of Community Cohesion in Small-Scale Fisheries. *International Journal of the Commons, 12*(1), 519–547.

Andino, J. (2009). *Spiny Lobster Initiative Report on Common Ground and Action Plans*. Spiny Lobster Initiative in Honduras.

Ansell, C., & Gash, A. (2007). Collaborative Governance in Theory and Practice. *Journal of Public Administration Research and Theory, 18*(4), 543–571.

Axelrod, R. (1984). *The Evolution of Cooperation*. New York, NY: Basic Books.

Barnes, M. L., Lynham, J., Kalberg, K., & Leung, P. (2016). Social Networks and Environmental Outcomes. *Proceedings of the National Academy of Sciences of the United States of America, 113*(23), 6466–6471.

Barnes-Mauthe, M., Arita, S., Allen, S. D., Gray, S. A., & Leung, P. (2013). The Influence of Ethnic Diversity on Social Network Structure in a Common-Pool Resource System: Implications for Collaborative Management. *Ecology and Society, 18*(1), 23.

Baumgartner, F. R., & Jones, B. D. (1991). Agenda Dynamics and Policy Subsystems. *The Journal of Politics, 53*(04), 1044.

Berardo, R., & Scholz, J. T. (2010). Self-Organizing Policy Networks: Risk, Partner Selection, and Cooperation in Estuaries. *American Journal of Political Science, 54*(3), 632–649.

Block, P. (2015). Reciprocity, Transitivity, and the Mysterious Three-Cycle. *Social Networks, 40*, 163–173.

Bodin, Ö., & Crona, B. I. (2009). The Role of Social Networks in Natural Resource Governance: What Relational Patterns Make a Difference? *Global Environmental Change, 19*(3), 366–374.

Boyd, R., & Richerson, P. J. (1987). The Evolution of Ethnic Markers. *Cultural Anthropology, 2*(1), 65–79.

Boyd, R., & Richerson, P. J. (2009). Culture and the Evolution of Human Cooperation. *Philosophical Transactions of the Royal Society B, 364*, 3281–3288.

Burt, R., (2005). *Brokerage and Closure: An Introduction to Social Capital.* Oxford, UK: Oxford University Press.

Van De Bunt, G., Van Duijn, M., & Snijders, T. (1999). Friendship Networks Through Time: An Actor-Oriented Dynamic Statistical Network Model. *Computational and Mathematical Organization Theory, 5*(2), 167–192.

Calanni, J. C., Siddiki, S. N., Weible, C. M., & Leach, W. D. (2014). Explaining Coordination in Collaborative Partnerships and Clarifying the Scope of the Belief Homophily Hypothesis. *Journal of Public Administration Research and Theory, 25*(3), 901–927.

Carpenter, D. P., Esterling, K. M., & Lazer, D. M. J. (2004). Friends, Brokers, and Transitivity: Who Informs Whom in Washington Politics? *The Journal of Politics, 66*(1), 224–246.

Centola, D., Gonza, J. C., & Egui, M. (2007). Homophily, Cultural Drift, and the Co-Evolution of Cultural Groups. *Journal of Conflict Resolution, 51*(6), 905–929.

Christie, P. (2014). *GFISH Project Case Study.* Unpublished—internal report.

Coleman, J. S. (1988). Social Capital in the Creation of Human Capital. *American Journal of Sociology, 94*, 95–120.

Cross, R., Borgatti, S. P., & Parker, A. (2002). Making Invisible Work Visible: Using Social Network Analysis to Support Strategic Collaboration. *California Management Review, 44*(2), 25–46.

Emerson, K., Nabatchi, T., & Balogh, S. (2012). An Integrative Framework for Collaborative Governance. *Journal of Public Administration Research and Theory, 22*(1), 1–30.

Faust, K. (2010). A Puzzle Concerning Triads in Social Networks: Graph Constraints and the Triad Census. *Social Networks, 32*(3), 221–233.

Fischer, M., Angst, M., & Maag, S. (2019). Co-Participation in the Swiss Water Forum Network. *International Journal of Water Resources Development, 35*(3), 446–464.

Freeman, L. C. (1978). Segregation in Social Networks. *Sociological Methods and Research, 6*(4), 411–429.

Gallemore, C. T., Prasti, H. D., & Moeliono, M. (2014). Discursive Barriers and Cross-Scale Forest Governance in Central Kalimantan, Indonesia. *Ecology And Society, 19*(2), 18.

Gerber, E., Douglas, A., & Lubell, M. (2013). Political Homophily and Collaboration in Regional Planning Networks. *Midwest Political Science Association, 57*(3), 598–610.

Geys, B. & Murdoch, Z. (2010). Measuring the 'bridging' versus 'Bonding' Nature of Social Networks: A Proposal for Integrating Existing Measures. *Sociology, 44*(3), 523–40.

Hagen, G., Killinger, D. K., & Streeter, R. B. (1997). An Analysis of Communication Networks Among Tampa Bay Economic Development Organizations. *Connections, 20*(2), 13–22

Henrich, J., & Boyd, R. (1998). The Evolution of Conformist Transmission and the Emergence of Between-Group Differences. *Evolution and Human Behavior, 19*, 215–241.

Henry, A. D. (2017). Network Segregation and Policy Learning. In J. Victor, A. Montgomery, & M. Lubell (Eds.), *The Oxford Handbook of Political Networks* (pp. 1–32). Oxford: Oxford University Press.

Henry, A. D., Lubell, M., & McCoy, M. (2010). Belief Systems and Social Capital as Drivers of Policy Network Structure: The Case of California Regional Planning. *Journal of Public Administration Research and Theory, 21*(3), 419–444.

Jackson, M. O. (2014). Networks in the Understanding of Economic Behaviors. *Journal of Economic Perspectives, 28*(4), 3–22.

Jasny, L., Waggle, J., & Fisher, D. R. (2015). An Empirical Examination of Echo Chambers in US Climate Policy Networks. *Nature Climate Change, 5*(8), 782–786.

Knoke, D. (1990). Networks of Political Action: Toward Theory Construction. *Social Forces, 68*(4), 1041.

Kossinets, G., & Watts, D. J. (2009). Origins of Homophily in an Evolving Social Network. *American Journal of Sociology, 115*(2), 405–450.

Krackhardt, D. (1998). Simmelian Ties: Super Strong and Sticky. In R. Kramer & M. Neale (Eds.), *In Power and Influence in Organizations* (pp. 21–38). Thousand Oaks, CA: Sage.

Kurian, M., Portney, K. E., Rappold, G., Hannibal, B., & Gebrechorkos, S. H. (2018). Governance of Water-Energy-Food Nexus: A Social Network Analysis Approach to Understanding Agency Behaviour. In S. Hülsmann & R. Ardakanian (Eds.), *Managing Water, Soil and Waste Resources to Achieve Sustainable Development Goals: Monitoring and Implementation of Integrated Resources Management* (pp. 125–147). Cham: Springer International Publishing.

Laird-Benner, W., & Ingram, H. (2010). Sonoran Desert Network Weavers: Surprising Environmental Successes on the U.S./Mexico Border. *Environment, 53*(1), 6–16.

Lazarsfeld, P. F., & Merton, R. K. (1954). *Friendship as a Social Process: A Substantive and Methodological Analysis, Freedom and Control in Modern Society* (pp. 18–66). New York: Van Nostrand.

Leifeld, P., & Schneider, V. (2012). Information Exchange in Policy Networks. *American Journal of Political Science, 56*(3), 731–744.

Lienert, J., Schnetzer, F., & Ingold, K. (2013). Stakeholder Analysis Combined with Social Network Analysis Provides Fi Ne-Grained Insights into Water Infrastructure Planning Processes. *Journal of Environmental Management, 125*, 134–148.

Lubell, M. (2007). Familiarity Breeds Trust: Collective Action in a Policy Domain. *The Journal of Politics, 69*(1), 237–250.

Lubell, M., & Lippert, L. (2011). Integrated Regional Water Management: A Study of Collaboration or Water Politics-as-Usual in California, USA. *International Review of Administrative Sciences, 77*(1), 76–100.

Lubell, M., & Niles, M. T. (2019). The Limits of Capacity Building. *Nature Climate Change, 9*(8), 578–579.

Lubell, M., Robins, G., & Wang, P. (2014). Network Structure and Institutional Complexity in an Ecology of Water Management Games. *Ecology and Society, 19*(4), 23.

Maag, S., & Fischer, M. (2018). Why Government, Interest Groups, and Research Coordinate: The Different Purposes of Forums. *Society and Natural Resources, 31*(11), 1248–1265.

Mcpherson, M., Smith-lovin, L., & Cook, J. M. (2001). Birds of a Feather: Homophily in Social Networks. *Annual Review of Sociology, 27*, 415–444.

Nowak, M. A. (2006). Five Rules for the Evolution of Cooperation. *Science, 314*(5805), 1560–1563.

Pons, P., & Latapy, M. (2005). Computing Communities in Large Networks Using Random Walks. In P. Yolum, T. Güngör, F. Gürgen, & C. Özturan (Eds.), *Computer and Information Sciences—ISCIS 2005* (pp. 284–293). Heidelberg: Springer Berlin Heidelberg.

Prell, C., Hubacek, K., & Reed, M. (2009). Stakeholder Analysis and Social Network Analysis in Natural Resource Management. *Society and Natural Resources, 22*(6), 501–518.

Provan, K. G., & Kenis, P. (2008). Modes of Network Governance: Structure, Management, and Effectiveness. *Journal of Public Administration Research and Theory, 18*(2), 229–252.

Putnam, R. (2001). Social Capital: Measurement and Consequences. *Canadian Journal of Policy Research, 2*(1), 11–17.

Ripley, R. M., Boitmanis, K. (2010). RSiena. http://www.stats.ox.ac.uk/?snijders/siena/.

Ripley, R. M., Snijders, T. A. B., Boda, Z., Vörös, A., & Preciado, P. (2017). *Manual for RSiena (2007)*. Oxford: University of Oxford.

Rudnick, J., Niles, M., Lubell, M., & Cramer, L. (2019). A Comparative Analysis of Governance and Leadership in Agricultural Development Policy Networks. *World Development, 117*, 112–126.

Sabatier, P. A., & Jenkins-Smith, H. C. (1993). *Policy Change and Learning: An Advocacy Coalition Approach*. Boulder, CO: Westview Press.

Smaldino, P. E., & Lubell, M. (2011). An Institutional Mechanism for Assortment in an Ecology of Games. *PLoS ONE, 6*(8), 1–7.

Smith-Lovin, L., and Mcpherson, J. M. (1987). "Homophily in Voluntary Organizations: Status Distance and the Composition of Face-to-Face Groups. *American Sociological Review, 52*(3), 370–379.

Snijders, T. A. B., van de Bunt, G. G., & Steglich, C. E. G. (2010). Introduction to Stochastic Actor-Based Models for Network Dynamics. *Social Networks, 32*(1), 44–60.

Turk, H. (1977). Organizations in Modern Life. San Francisco: Jossey-Bass.

Ulibarri, N., & Scott, T. A. (2017). Linking Network Structure to Collaborative Governance. *Journal of Public Administration Research and Theory, 27*(1), 163–181.

Weible, C. M., & Sabatier, P. A. (2005). Comparing Policy Networks: Marine Protected Areas in California. *Policy Studies Journal, 33*(2), 181–201.

4

Network Embeddedness and the Rate of Water Cooperation and Conflict

James Hollway

Introduction

Managing water resources across borders of any scale is challenging (Lubell 2013; Ingold et al. 2016), but international basins present a special challenge. Comprehensive governance of international basins is rare (Wolf et al. 2003; Conca 2005) and many practitioners and scholars remain concerned that demographic dynamics, agricultural pressures, and climate change (Fischhendler 2004; Tir and Stinnett 2012) may make international "water wars" more common in the future (Hensel and Brochmann 2009). While studies have repeatedly found that water-related cooperation is more common than conflict (Wolf et al. 2003; Kalbhenn 2011), many country dyads do slip into water-related conflict. This chapter asks: why is cooperation more frequent than conflict?

To date, the literature on international water management has focused on three main areas: the establishment of international water agreements

J. Hollway (✉)
Graduate Institute of International and Development Studies, Geneva, Switzerland
e-mail: james.hollway@graduateinstitute.ch

© The Author(s) 2020
M. Fischer, K. Ingold (eds.), *Networks in Water Governance*, Palgrave Studies in Water Governance: Policy and Practice, https://doi.org/10.1007/978-3-030-46769-2_4

or organizations (e.g. Dinar et al. 2011; Zawahri and Mitchell 2011; Tir and Stinnett 2012); the relationship of freshwater scarcity to militarized interstate disputes (e.g. Furlong et al. 2006; Gleditsch et al. 2006); and the frequency of water cooperation or conflict events (e.g. Yoffe et al. 2003; Hensel and Brochmann 2009; Kalbhenn 2011; Bernauer and Böhmelt 2014). This latter "basin at risk" literature is particularly advanced in collecting, coding, and analyzing date-stamped data on water-related cooperative and conflictual events between countries.

However, though these literatures regularly employ statistical models in addition to case studies (Wolf et al. 2003; Zeitoun and Mirumachi 2008; Brochmann and Gleditsch 2012), there are few applications of network models (Berardo and Gerlak 2012, being a rare exception). While statistical network models are increasingly used to study water policy networks (e.g. Ingold et al. 2016) and complex networks of international institutions in other environmental fields (Biermann et al. 2020), the author is not aware of any that model water events as a network. This is lamentable, since three central dependencies in these "basin at risk" datasets lend themselves to network theories and methods. First, cooperation and conflict are not mutually exclusive (Zeitoun and Mirumachi 2008), as sometimes treated in this literature, but may prompt or suppress the other. Second, states' cooperation and conflict over water resources are often public, which may allow states to condition their behavior on the behavior of others that they observe. Third, because these events are date-stamped, there is information about the sequencing and, indeed, timing of cooperation and conflict within and across dyads that can be exploited to support inference about not only with whom states cooperate or come into conflict but also when.

This chapter demonstrates that network mechanisms can help explain why some states act cooperatively and conflictually more often than others as well as with whom they cooperate or come into conflict. While the social networks literature to date has been more interested in the latter, regarding actors' choices, we should also begin to explore the corollaries most network mechanisms have for the rate of network activity of different types. The chapter argues, for instance, that a state's embeddedness in triangles of local cooperative or conflictual behavior affects its rate of cooperation and conflict.

This chapter makes four main contributions. First, it complements other chapters in this volume by offering an example of applying social network theory and models to study *international* water cooperation and conflict. Second, it offers a first application of statistical *network* models to international water events. It models international water events as network events using dynamic network actor models (DyNAM; Stadtfeld et al. 2017a) to model not only the location but also the timing of cooperation and conflict events. Third, it demonstrates for the first time the use of DyNAMs for coevolving, signed networks. Fourth, it represents one of the first empirical emphases by an actor-oriented network model of the rate rather than choice part of the model.

The rest of the chapter is structured as follows. The next section outlines key expectations from network theory about where and when cooperation and conflict should take place, and summarizes typical theoretical expectations about international water cooperation and conflict from the literature on political geography, political economy, and political institutions. The following section describes the International Rivers Cooperation and Conflict (IRCC) water event data used here. Next, I introduce the DyNAM model and briefly explain how coevolving signed DyNAMs can be modeled. The penultimate section presents and interprets the results obtained by fitting this model to both water cooperation and conflict events. Finally, I conclude by reflecting on the main findings and their generalizability, the practical policy advice that can be drawn from them, and potential next steps for scholarship in the area.

Theory

This section introduces the insights political networks can offer on what makes international water cooperation more frequent than conflict, before recounting typical expectations currently highlighted in three main literatures related to water events: political geography, political economy, and political institutions.

Theories of Political Networks

The "basins at risk" literature conceives of cooperation and conflict as events that occur on a particular date from one state to another. Although these events are associated with particular date-stamps, they have more enduring salience, persisting in the memory of the actors that experienced them and, when public, beyond. As these events accumulate, they can be conceived of as constructing a network of events between actors that structures and informs when and where future events occur. This is important, since these events are not independent but cluster in dyads and triadic configurations. Political networks encourage us to not only account for such clustering, which would otherwise lead to underestimated standard errors, but also associate such configurations with endogenous processes and mechanisms of interest. As Soliev et al. (2017, p. 148) argue, "network effects […] form the so-called 'baggage' in riparian relationships". Such "baggage" can slow or accelerate further cooperation and conflict. In this chapter, I outline three basic network configurations, oriented around monads, dyads, and triads, and outline expectations for how they affect both cooperative and conflictual timing (rate) and location (choice). This paper emphasizes the third set of effects as most illustrative of a network approach and most interesting for water management.

First, cooperation and conflict tend to follow past cooperation and conflict. Actors regularly repeat past events, establishing well-worn patterns (Uzzi and Lancaster 2003): we would expect a cooperative actor to continue cooperating (and, perhaps, avoid conflict) and an actor that has been in conflict recently to repeat this (and avoid cooperation). This activity effect is outlined in Fig. 4.1(a), where the dashed line represents a new event and the solid lines the recent events. We would also expect repetition in a state's choice of cooperation or conflict partner. In the international water management literature, this has been operationalized as "peace history" (e.g. Brochmann and Gleditsch 2012), but here we measure this as the entrainment of past cooperation and conflict on recent behavior (see Fig. 4.1(d)).

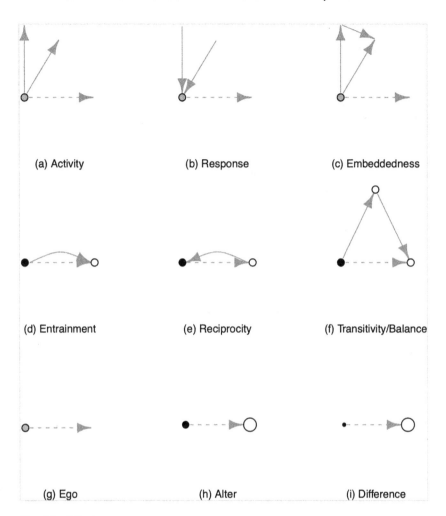

Fig. 4.1 Effects

Second, incoming network ties are also important for when and with whom states cooperate or come into conflict. Directed events demand a response (Fig. 4.1(b)) from the recipient actor while the event is still salient, though not necessarily in kind. Being on the receiving end of conflictual behavior may demand a cooperative response, if not with the sender then with others. Where the target chosen is specified as the sender of a previous event, we speak of reciprocation (Fig. 4.1(e)). Failing to respond directly would be to implicitly accept status inferiority (Gould 2002, p. 1151). Wolf (1997) highlights how a lack of recognition in Palestine and Kurdish examples blocked cooperation. But we might also expect events to be exchanged, with actors reciprocating conflict with cooperation, as they seek to settle issues. For example, responding to a conflict-inducing action with a timely cooperative move, such as information-sharing or financing, can defuse the situation and restore cooperation (Wolf 1997, p. 350).

Perhaps the classic social networks dependencies, however, are those that involve triadic configurations where an actor's partners are themselves connected. These are most commonly elaborated in the context of partner choice (transitivity, Fig. 4.1(f)): we are more likely to befriend a friend's friend, for example (see Granovetter 1985, p. 490). Since the current network is signed, including both positive (cooperative) and negative (conflictual) events or ties, "structural balance theory" may also be applicable (Cartwright and Harary 1956). This theory argues that unbalanced configurations, such as being in conflict with a cooperative partner's other partner, induces cognitive dissonance for the actors involved that demands resolution through, for example, cooperating with this other partner or expanding the conflict. Third parties can support the restoration of a cooperative relationship by potentially brokering the resolution of any disagreements (Wolf 1997, p. 350; Simmel 1950). We would thus expect balanced configurations to be more likely than unbalanced configurations:

H1 *Actors are more likely to cooperate with cooperative partners' cooperative partners*

H2 *Actors are less likely to be in conflict with cooperative partners' cooperative partners*

H3 *Actors are more likely to cooperate with conflict partners' conflict partners*

H4 *Actors are less likely to be in conflict with conflict partners' conflict partners*

Expectations for the application of triadic configurations on rate are less well elaborated, at least directly. Granovetter (1985) argued that actors' embeddedness in their local networks affected how they perceived and acted within the network. Sets of cooperative partners can reinforce cooperative norms and sets of conflict partners may reinforce norms of conflict too. Therefore, one might argue that the more an actor is embedded in cooperative triads (Fig. 4.1(c)), the more it will cooperate, and the same with conflict. This can be contrasted with structural holes theory, in which Burt (2004) argues that those who are less embedded are freer to exploit opportunities in the network afforded by their brokerage positions, and consequently act more often. We would thus expect the relationship between embeddedness and rate to be inverted. Here I outline the main expectations of embeddedness:

H5 *Actors are more likely to cooperate when embedded in recent cooperative triads*

H6 *Actors are less likely to act conflictually when embedded in recent cooperative triads*

H7 *Actors are less likely to cooperate when embedded in recent conflictual triads*

H8 *Actors are more likely to act conflictually when embedded in recent conflictual triads*

These eight expectations relate triadic configurations of past cooperation and conflict to the timing (rate) and location (choice) of further cooperation and conflict and represent the main hypotheses investigated in this chapter. To support identification of these network effects however

requires that we also control for common explanations in the three literatures that have treated water-related cooperation and conflict to date.

Theories of Political Geography

A common factor expected to provoke or ameliorate conflict is the availability or scarcity of water. This is in line with a neo-Malthusian perspective that expects resource scarcity to provoke conflictual behavior (Hensel
and Brochmann 2009). Zawahri and Mitchell (2011) find that greater
dependence on cross-border freshwater resources makes cooperation
more likely, while higher precipitation levels make it less likely. I therefore
expect water availability or scarcity to drive both cooperation and conflict.

Another factor is the dependency of a downstream state on an upstream
state for appropriate water quantity and quality (Mitchell and Keilbach
2001). Scholars have put considerable effort into measurement here
(Furlong et al. 2006; Gleditsch et al. 2006; Beck et al. 2014), perhaps
driven by mixed results. Furlong et al. (2006) and Gleditsch et al. (2006)
were unable to distinguish whether upstream/downstream geography
impacted militarized interstate disputes, Dinar et al. (2011) found that
the riparian configuration was significant in only part of the estimates,
and Munia et al. (2016) found no direct relationship between upstream
water use and the number of conflictive and cooperative events.
Brochmann and Gleditsch (2012) argue that any specific riparian relationship simply confounds the overwhelming effect of contiguity on the
frequency of interstate relations, conflictual or cooperative, noting that
only 17 contiguous dyads do not share a river. I therefore expect no relationship for water dependency, but for contiguity.

Theories of Political Economy

An abiding expectation for interstate cooperation is that democratic
countries behave more cooperatively. A neo-Kantian perspective main-
tains that democracies cooperate more together (Mansfield et al. 2002),
and Brochmann and Gleditsch (2012) find that political regime type

significantly affects water cooperation and conflict. But democracies also better govern water resources internally, leading to fewer internal water-related conflicts that can spill out (Wolf 1997).

More developed countries are also expected to be more cooperative. Like democracies, developed countries may have better governance and the capacity necessary to resolve conflicts. Dinar et al. (2011) find that more developed states are in a position to provide incentives, such as financial transfers, to less-developed states so as to facilitate an international agreement. But developed countries may also have access to alternative sources of water to mitigate water dependency, and Wolf (1997) argues that different levels of development can exacerbate conflict.

Theories of Political Institutions

While there is currently no overarching water convention (Dellapenna and Gupta 2008)—though the United Nations Watercourses Convention of 1997 has been in force since 2014, many key riparian states have not ratified or acceded—there are hundreds of bilateral and multilateral water agreements currently in place (Zawahri and Mitchell 2011). The literature on institutional design and effectiveness in International Relations have classified a range of institutional features (see Koremenos et al. 2001), five being most common in the literature on water cooperation (Mitchell and Keilbach 2001; Berardo and Gerlak 2012; Tir and Stinnett 2012): delegation, allocation, enforcement, dispute resolution, and flexibility.

Some riparian states have delegated governance functions to regional basin organizations (RBOs) (Wolf 1997). RBOs' secretariats play various roles that can help states absorb stresses from competing water uses. Secretariats can reduce transaction costs to further cooperation (Wolf 1997; Zawahri and Mitchell 2011) through what Schmeier and Shubber (2018) call "institutional anchoring". For example, the Mekong River Commission Secretariat has been key in mitigating conflicts around parties' infrastructure projects (Schmeier et al. 2015).

Water institutions also vary in how explicit and clear water allocation rules are, though the effect is not as clear. On the one hand, unclear or contested terms have been found to lead to conflict (Hansen et al. 2008), and clear allocation rules should mitigate disputes since there is less space for debate. On the other hand, clear allocation rules can also constrain parties leaving conflict the only recourse. Rayner et al. (2005) argue that while water managers' prefer highly specified institutionalized systems to ensure reliable water access under typical scenarios, these systems can falter when challenges, such as flow variability, occur. Though managers usually seek cooperation in response, unreciprocated cooperative moves can lead to blame, disputes, and conflict.

Strong enforcement mechanisms are generally thought to consolidate cooperation and stave off conflict. Institutions can consolidate cooperation by enforcing a pattern of cooperation that helps preclude disputes (Wolf 1997, pp. 349–350), but Hansen et al. (2008) argue this depends on the enforcement capabilities of the institution itself. One challenge with all these institutional features, however, is their political feasibility (Fischhendler 2004). Instituting cooperation with strong enforcement mechanisms may not be possible where it is needed, and instituted where it is not.

Institutions endowed with dispute resolution processes are also thought to facilitate cooperation and conflict resolution. Mechanisms to settle disputes vary, from binding arbitration or adjudication to non-binding mediation, though in practice, many are "innocuous", requiring little more than meetings (Wolf 1997). Still, the ability to even facilitate agreement over scientific data can have important ramifications for a conflict's resolution. Hensel and Brochmann (2009) find that, although river agreements do not prevent conflict, they provide a starting point for negotiations over disputed river claims and can more speedily return a relationship to a more cooperative setting.

Lastly, designing flexible institutions can support cooperation in the face of conflict. Since one stressor in riparian relationships is fluctuation in resource availability, institutions that can adapt to changing circumstances will be more resilient (Yoffe et al. 2003). Fischhendler (2008) discusses the utility of ambiguities left in the original arrangement to allow for flexibility as problems and preferences change. Though it can lead to protracted disagreement, Fischhendler argues that institutional adaptations do not happen in a vacuum, but depend on the roles and preferences of the actors around the institution.

Case

To identify lessons on water cooperation and conflict that might generalize to international river basins around the world, scholars have sought to complement existing case studies with the analysis of datasets that record interaction *events* between countries (Zeitoun and Mirumachi 2008). Event databases have a long history in International Relations. Originating in the early 1960s, event databases scrape news media sources for day-to-day interstate interactions, and then manually or automatically code them to some scale of cooperation and conflict.

International water governance has seen some of the most extensive and targeted efforts in this area, certainly more so than in other environmental fields. Here I consider two of the most recently developed datasets, the Transboundary Freshwater Disputes Water Events Database (TFDD; Yoffe et al. 2003; Wolf et al. 2003) and the International Rivers Cooperation and Conflict event database (IRCC; Kalbhenn 2011; Kalbhenn and Bernauer 2012; Bernauer and Böhmelt 2014). Unlike earlier efforts, both collect both cooperative and conflictual events and are specifically water-related. This issue focus and type scope enables more complete, precise inference on international interactions.

This chapter uses the IRCC data for two main reasons. First, the IRCC data are transparently coded from a more homogenous set of sources. Though the TFDD offers data for a longer time period (1948–2008 compared to 1997–2007), as Bernauer and Böhmelt (2014, p. 121) explain, "major changes in the availability of news media texts over time (notably the advent of the digital revolution) make it problematic to use event data coded from partly changing sources for a very long period of time". In any case, despite the shorter time frame, the IRCC dataset includes more of certain types of events. Second, many water-related events, whether statements or actions, are directed. The TFDD does not code the direction of events, but the IRCC does. However, Kalbhenn and Bernauer (2012) suggest that "[d]isaggregating the data to monthly, weekly, or even daily events makes little sense in our context because most covariates commonly used in this area of research (e.g. economic indicators, political system data) are only available on a yearly basis". But while indeed GDP is only recorded annually, the Polity dataset offers a date-stamped record of changes to countries' level of democracy or autocracy. Moreover, if the events are date-stamped, then we can make more precise inference about the sequencing of events between different actors and of different types. I thus use the date-stamped IRCC event data.

The data used in this chapter are thus all the water events with date-stamps in the IRCC database. Figure 4.2 plots the distribution of events according to their IRCC score, which ranges between −6 (most conflictive, i.e. violent interstate dispute with declaration of war) and +6 (most cooperative, i.e. ratification of freshwater treaty) (see Kalbhenn and Bernauer 2012, for more details). In practice though, relatively few events were coded beyond 3 in absolute value. At the extremes are, for example, Israeli air raids that targeted an area being excavated as part of the Al-Asi Dam project on the Lebanese-Syrian border (coded −5), or India and Bangladesh signing an agreement to share the water of the Tista and six other rivers (coded 5).

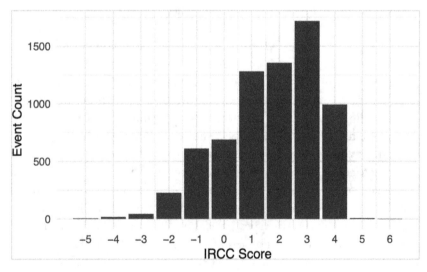

Fig. 4.2 IRCC water cooperation and conflict events

Figure 4.2 shows a left skewed distribution, corresponding to the observation that water cooperation is more common than conflict. Since the middle of the range is less distinctive (the difference between −1 and −2, for instance, is whether a statement is "mild" or "strong"), I follow previous work in binarizing this distribution into "conflict" and "coop-eration" events, which also aids in relating results to this literature. All events with an IRCC score more than 0 were classified as cooperative, and all events with an IRCC score less than 0 were classified as conflic-tual. Given the rarity of "water wars", this category can perhaps better be categorized as political *disputes*, but regardless of label, these negatively signed events are worth examining separately from cooperation. As Zeitoun and Warner (2006, p. 437) state: "the absence of war does not mean the absence of conflict". Those events with an IRCC score of exactly 0 were coded as a third, neutral category and not modeled here. This resulted in a total of 908 conflictual events and 5360 cooperative events.

Here I define an event as a date-stamped action from a sender to a receiver. Agreements are defined as two directed actions, one each way. There is an important duality here though: these actions are instanta-neous (within the continuous-time assumptions of the model) but also define the starting point of a tie through the residue such a tie creates.

This data comprises 104 states that sent or received at least one coop-erative or conflictual water event in the period in question (1997–2007) as the nodes of the network. There are thus 1.0712×10^{4} potential dyads in the data, though many of these are empty since water cooperation and conflict is largely spatially local. It is, however, not exclusively spatial and so tie opportunities were not constrained to contiguous dyads, despite the option being available in the most recent version of goldfish, the soft-ware used. This is later validated by the absence of strong contiguity effects.

One way to explore how dyadic relationships, defined as chains of interactions, have progressed is as a sequence (for a recent introduction to sequence analysis, see Cornwell 2015). Figure 4.3 plots the trajectories of IRCC scores in each directed-dyad relationship (that is, India-Pakistan and Pakistan-India each receive a line). Two chief observations can be drawn from this plot. First, the density of lines toward the left hand part of the plot signal that many relationships are relatively short, though the

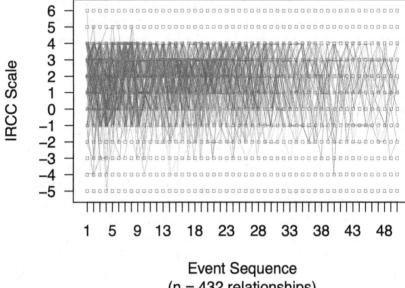

Fig. 4.3 Event sequences

thick lines extending half way across the plot also indicates that many relationships see a longer history of interactions. Indeed, while Fig. 4.3 only plots relationships up to 50 interactions to improve readability, several relationships had much longer chains of interactions. For example, in this period, Hungary sent 238 events (cooperative and conflictual) to Slovakia, and Slovakia reciprocated by sending 173 events. Similarly, Romania and Ukraine sent each other 215 and 165 events, respectively. In all, 21 directed dyads have chains of events during this period that are longer than 50 events.

Second, the line density in the middle of the graph between −1 and 4 accords with 1 and the finding that water wars are rather rare. But it also shows that most relationships over a number of events experience cooperative and, at some point, conflictual or neutral events. Cooperative relationships rarely stay cooperative; nor are conflictual relationships consigned to remain conflictual. Indeed, it is possible to see a common sequence early in the relationship as shown by where the lines are thickest. Relationships seem to often start cooperatively, then fall into some conflict by the third or fourth event, before returning to more cooperative relations. As De Stefano et al. (2010, p. 873) note, while a series of events may pass through several conflictive intensities over time, the process does not necessarily evolve linearly.

Such event data does have its caveats (see Schrodt 2012). First, many international events often go unreported, because they either are not deemed newsworthy or are kept hidden for strategic reasons. Second, popular media often presents a biased record of events, generally favoring the country in which they are based. Third, the data quantity that can be collected can introduce sensitivities relating to coding rules. It is therefore important that these are as transparent as possible. Overall though, event data can serve as an efficient trace of cooperative and conflictual relations between states, offering an improvement in granularity and the avoidance of some biases over other types of data often used.

Methods

The political science literature on international water cooperation and conflict has taken two main methodological approaches. Perhaps the most common approach remains the case study (e.g. Bréthaut 2016; Verweij 2017). Case studies can offer a rich account of specific interstate water relationships, but are said to struggle with generalization, despite a few comparative efforts (e.g. Knieper and Pahl-Wostl 2016). The principal alternative is the growing number of econometric studies of water event data (Furlong et al. 2006; Gleditsch et al. 2006; Hensel and Brochmann 2009). Yet, there have been remarkably few works that explicitly look at temporal dependencies in such data let alone structural dependencies.

Statistical network models offer various ways to not only account for but also explicitly explore structural dependencies (Lubell et al. 2012). Classic network models include exponential random graph models (ERGMs; Lusher et al. 2013) and stochastic actor-oriented models (SAOMs; Snijders et al. 2010). There are important differences relating to whether they are tie-based or actor-oriented and how they treat time (see Block et al. 2016, 2018), but neither are really equipped to fully leverage date-stamped tie data (events) because they explore dependencies among tie observations by simulating the most likely series of tie changes that lead to network structures and dispense with any information about the order of ties/events.

Two other statistical network models are better equipped, however: relational event models (REMs; Butts 2008) and dynamic network actor models (DyNAMs; Stadtfeld et al. 2017a). Both ultimately model the rate at which we expect to see ties in particular configurations (readers are referred to Stadtfeld et al. 2017b, for more details). The chief distinction between them is that, as an actor-oriented model similar to SAOMs, DyNAMs separate the overall tie rate into two functions as shown in equation 1: a Poisson process governing the rate at which actors make ties, and a multinomial choice model that, given a particular actor chosen to make a tie (i), governs which other node she chooses (j rather than any other node from among the set of A others). Each function can be specified with statistics (s or t) that capture salient, current features of the network, such as nodal attributes or structural configurations. How these are weighted by parameters θ (for the rate function) and β (for the choice function) determine actors' competing rates and competing attractiveness as a recipient for a tie, respectively.

$$\lambda_{ij}^{\mathrm{DyNAM}}(x,\theta,\beta,s,t,A) = \underbrace{\exp(\theta^T s(x,i))}_{\mathrm{Rate}}\underbrace{\frac{\exp(\beta^T t(x,i,j))}{\sum_{k\in A}\exp(\beta^T t(x,i,k))}}_{\mathrm{Choice}}$$

This two component structure suggests a literal interpretation that an actor first becomes active and then decides to which other node to send a tie. However, this is not a necessary interpretation. Just as a choice function need not be interpreted literally as actors operating under strict and explicit utility maximization rules, but as capturing how a concatenation of different factors conjoin to make some ties more likely choices than others, so too can the separation be seen as largely artificial as a way to allow researchers more flexibility in specifying models and to allow them to interpret timing and choice separately.

DyNAMs have three main advantages: precision, performance, and properties. First, because they use information about the order and timing of events, they offer greater precision than SAOMs, and because they allow a flexible specification of rate and choice, they also offer greater precision than REMs. Second, because they model information about tie ordering directly, they can forego the simulation SAOMs and ERGMs rely on, and because they separate tie rates into actor rates and choice, they also involve a lower order of computational complexity than REMs. Details about the estimation are provided in Stadtfeld et al. (2017a, b) and Stadtfeld and Block (2017) illustrate the comparison with REMs in particular. Lastly, DyNAMs allow a plethora of new effects that leverage information contained in events' timing. This chapter demonstrates two of them: windowed effects that only count configurations within a specific temporal window, and weighted effects that depend on how many events have been sent.

This chapter also demonstrates how signed networks (discussed in Stadtfeld et al. 2017a) can be modeled as coevolving directed networks (introduced in Stadtfeld and Block 2017) to explore dependencies between positive and negative valence ties. As described in Stadtfeld et al. (2017a, pp. 17–18), signed networks can be modeled as dependent subprocesses, using effects that capture structural configurations relating to one network in the model specification for the other network to model dependencies between them. It also represents one of the first empirical studies that fully leverage the flexibility of the rate function for exploring variation in the rate of actors' activity.

The main effects have already been laid out in Fig. 4.1. In addition to activity, response, embeddedness, entrainment, reciprocity, and transitivity/balance are three further types of effects, ego (Fig. 4.1(g)), alter (Fig. 4.1(h)), and difference (Fig. 4.1(i)), that are used to map the effects of political geographic, economic, and institutional variables on actors' rate and choices in the dynamic network of cooperative and conflictual events. The ego effects capture the effect of water availability, a state's economic size, or its regime on cooperative or conflictual activity. The alter effects capture the effect these variables have on a state being selected as the recipient of a cooperative or conflictual event. And the difference effects help us investigate whether states are selected as recipients because they are dissimilar to the sender in these variables. Lastly, contiguity and water dependence are included as additional networks that are expected to entrain cooperation and conflict. The weighted versions and a one-year window were used for all structural effects to capture all recent events. The next section presents the results of fitting a DyNAM specified with these effects.

Analysis

Dynamic network actor-oriented models (DyNAMs), including both rate and choice model types, were fitted to conflict and cooperation events drawn from the IRCC dataset using the goldfish package version 1.4.0 "Bristol Shubunkins". All model results presented here converged with a maximum absolute score below 0.001. Diagnostics (see Hollway and Stadtfeld 2017, for more details) suggest little temporal heterogeneity in the models and few outliers. The final results are presented in Table 4.1. Robustness checks included the presence of neutral events, various combinations of weighted and windowed versions of the main structural effects, and some additional variables present in the IRCC dataset such as shared basins without affecting the chapter's main conclusions.

I begin by interpreting the rate models. First, note that we can interpret the intercept here as the unconditional waiting time for a country to send an event. On average, countries send a water-related cooperative event every 39 days and a conflictual event every 72 days, reflecting how much more common water-related cooperation is than conflict (Wolf 1998).

Of particular interest here is how a country's (recent) local network of cooperative and conflictual events affects the frequency of cooperation and conflict. Activity was statistically significant and positive across the board: the more countries have cooperated or been in conflict in the last year, the more likely they are to both cooperate and be in conflict again. Only one response effect was statistically significant: incoming conflict behavior makes states less likely to cooperate (with any other country). The embeddedness effects were all statistically significant, however rather

Table 4.1 Results

Effect		Cooperation			Conflict	
		Est.	S.E.		Est.	S.E.
Rate		(N = 6540, LL −86117.11)			(N = 2360, LL −15835.18)	
Intercept		−15.037	(0.093)***		−15.638	(0.244)***
Coop activity		0.011	(0.003)***		0.014	(0.004)**
Conf activity		0.010	(0.002)***		0.042	(0.003)***
Coop response		0.004	(0.003)		0.007	(0.005)
Conf response		−0.008	(0.002)**		−0.003	(0.005)
Coop embeddedness	H5	0.061	(0.001)***	H6	0.041	(0.003)***
Conf embeddedness	H7	−0.009	(0.003)**	H8	−0.216	(0.030)***
Ego's water		−0.049	(0.003)***		−0.132	(0.010)***
Ego's regime		0.012	(0.003)***		0.017	(0.007)**
Ego's economy		−0.139	(0.012)***		−0.212	(0.031)***
Choice		(N = 5360, LL −15064.17)			(N = 908, LL −1881.1)	
Coop entrainment		0.120	(0.008)***		0.106	(0.022)***
Conf entrainment		−0.057	(0.009)***		0.083	(0.021)***
Coop reciprocity		0.071	(0.008)***		0.097	(0.023)***
Conf reciprocity		−0.056	(0.009)***		−0.073	(0.021)***
Coop balance	H1	0.659	(0.012)***	H2	0.805	(0.038)***
Conf balance	H3	0.001	(0.029)	H4	−0.003	(0.165)
Institutional delegation		0.044	(0.175)		−0.037	(0.401)
Institutional allocation		0.718	(0.072)***		1.258	(0.199)***
Institutional enforcement		−0.710	(0.103)***		−0.993	(0.413)*
Institutional resolution		0.388	(0.049)***		0.074	(0.144)
Institutional flexibility		1.157	(0.173)***		1.451	(0.382)***
Alter's water		−0.020	(0.004)***		−0.026	(0.015)
Water differences		−0.116	(0.005)***		−0.207	(0.018)***
Alter's regime		−0.012	(0.003)***		−0.012	(0.008)
Regime differences		−0.017	(0.004)***		0.070	(0.010)***
Alter's economy		−0.070	(0.015)***		0.024	(0.046)
Economy differences		−0.384	(0.020)***		−0.496	(0.063)***
Contiguity		−0.142	(0.093)		−0.716	(0.311)*
Water dependence		0.183	(0.077)*		−0.026	(0.248)

$^*p < 0.05$, $^{**}p < 0.01$, $^{***}p < 0.001$

than being normative as expected in H1–H4, being embedded in a cooperative triangle supports both cooperative and conflictual behavior, whereas being embedded in a conflictual triangle suppresses both types of behavior. For example, a state that has cooperated with a partner that has cooperated with another of its partners in the last year will cooperate 2 days faster than the baseline and act conflictually 3 days faster than the baseline. But a state that has been in conflict with states that were themselves in conflict will cooperate 1 day slower than the baseline and act conflictually 17 days slower than the baseline. This suggests that being embedded in cooperative triads emboldens actors, and being embedded in conflictual triads makes actors more cautious. I propose to call this facilitative embeddedness rather than the normative embeddedness outlined by Granovetter.

Next, the ego effects are all statistically significant and go in the same direction for both conflict and cooperation. Consistent with Dellapenna and Gupta (2008), Hensel and Brochmann (2009), and Zawahri and Mitchell (2011), countries that suffer from water scarcity are more active in cooperation and conflict. Whereas a country that receives the minimum rainfall observed in the data will cooperate every 44 days and be in conflict every 95 days, a country that receives maximum rainfall will only cooperate every 140 days or be in conflict every 6.17 years. This supports the general finding in the literature that water availability affects cooperation and especially conflict. Other ego effects suggest that poorer, democratic countries are both more cooperative and conflictual. Fully democratic countries cooperate over water every 35 days and are in conflict every 60 days compared to 44 and 85 days for fully autocratic countries. Poor countries cooperate every 81 days and are in conflict every 216 days compared to 162 and 624 days for rich countries. This somewhat counterintuitive result is probably driven by major riparian countries such as India, who are democratic, often in conflict with their neighbors, and may also often appear in the online media sources used in the IRCC data.

Network effects also affect with whom countries cooperate or come into conflict (choice). Results for entrainment, reciprocity, and balance are complicated and best read together. Countries cooperate with those with whom they have cooperated and that have cooperated with them in the last year, and avoid cooperation with those with whom there was conflict in the last year. Conflict appears to be preceded not only by past conflict with that country, but also cooperation, suggesting that close cooperation can create friction too. And while countries seem to be attracted to the balanced configuration of cooperating with a partner's partner, they are also attracted to the imbalanced configuration of being in conflict with a cooperative partner's cooperative partner. This adds more mixed evidence for the structural balance theory (Harrigan and Yap 2017). To sum up these configurations with an example, country a is most likely to cooperate with country b if, in the last year, a has cooperated and not come into conflict with b, b has cooperated and not come into conflict with a, and a has cooperated with c who has also cooperated with b. Country a is most likely to come into conflict with b if, in the last year, a has cooperated or come into conflict with b, but b has only cooperated and not acted conflictually toward a, and a has also cooperated with c who has also cooperated with b. Overall, this suggests a complex embedding of riparian relationships that drive both cooperation and conflict, as illustrated in Fig. 4.3, and that again cooperative embedding can create frictions that result in conflicts.

Note that this deepening of the relationship is net of typical geographical controls, such as contiguity and water dependency. Contra recent literature (Furlong et al. 2006; Gleditsch et al. 2006), water dependency is statistically significant, but only for cooperation and not conflict. Like Brochmann and Gleditsch (2012), contiguity correlates with conflict, but is unexpectedly negative. However, this effect needs to be interpreted in light of the (weighted) conflict and cooperation ties above that would already capture any repeated interaction among neighboring states: a country is unlikely to come into conflict over water with a neighboring country that it had not already cooperated or fought with in the past. Countries cooperate and come into conflict with those who have similar levels of water availability, and especially cooperate with those who are suffering from water scarcity. They also cooperate with similar regimes (especially if they are authoritarian) and come into conflict with different regimes. Lastly, they cooperate and conflict with similarly sized economies, and especially cooperate with smaller economies. This tendency toward smaller and authoritarian states is likely due to the presence of various types of water-related support, such as infrastructure investment, in the dataset.

Finally, several institutional features are important here too. Strong allocation and flexibility provisions prompt both cooperation and

conflict, whereas strong enforcement provisions suppress both cooperation and conflict. Strong resolution provisions also support cooperation
but not conflict. Delegation did not appear significant here, but did in
some of the robustness tests. Since institutional design features do multiple things, it is perhaps unsurprising that the results are ambiguous,
suggesting more work is needed here (Biermann et al. 2020).

Conclusions

This chapter has demonstrated how network theory and statistical network modeling can be applied to international water-related cooperation
and conflict event datasets. It also serves as first demonstrations of
coevolving signed DyNAMs and a fully specified and emphasized rate
function among actor-oriented network models.

The chapter has not only been demonstrative though. It has argued
that countries' cooperation and conflict is structured by the residue of
past events between them and with their network neighbors. Using
dynamic network actor-oriented models (DyNAMs), and controlling for
typical explanations in the literatures on water cooperation and conflict,
I find that network configurations do affect when and with whom countries act cooperatively and conflictually. Most interesting is that countries
that are embedded in cooperative relationships with two or more other
states act quicker, both cooperatively and, it seems, conflictually, but that
being embedded in conflictual relationships slows them down. I suggest
that cooperative embedding is *facilitative* and emboldens activity, whereas
actors that are embedded in conflictual relations exercise caution, but
further research is necessary to examine the effect of embeddedness on
rate in different settings.

A chief attraction of datasets like the IRCC for both scholars and practitioners is the promise of more generalizable findings (Bernauer and
Böhmelt 2014). A well-specified and well-performing statistical model
on carefully constructed and cleaned data that identifies average effects
for various policy-relevant mechanisms can inform future policy about
the likely effects of policy decisions. However, expectations must be managed for what can be predicted or forecasted when models (correctly)

incorporate temporal and structural endogeneities and dependencies. Forecasting beyond the immediate future with models that include significant network effects faces the challenge that these effects capture dependencies and endogeneities that can fork the system into paths with quite different contexts for action (Block et al. 2018).

Yet network models can still provide practical policy advice. Structural effects highlight dependencies that make our inferences about other effects less biased, but can also suggest social points of leverage on relationships. For example, recent tensions over Ethiopia's Grand Ethiopian Renaissance Dam highlight the role that third parties, particularly Sudan, can play in mediating and mitigating the conflict, though these results caution that riparian relationships are neither simple nor straightforward. International water institutions therefore need to be designed and resourced so that they can manage the parties, not the water, or what Van Ast (1999) calls "interactive water management". This points to the need for further networks research in the area, in ways that fully leverage the increasingly detailed data available but take the networked structure of states' interactions seriously.

References

Beck, L., Bernauer, T., Siegfried, T., & Böhmelt, T. (2014). Implications of Hydro-Political Dependency for International Water Cooperation and Conflict: Insights from New Data. *Political Geography, 42*(c), 23–33.

Berardo, R., & Gerlak, A. K. (2012). Conflict and Cooperation Along International Rivers: Crafting a Model Of Institutional Effectiveness. *Global Environmental Politics, 12*(1), 101–120.

Bernauer, T., & Böhmelt, T. (2014). Basins at Risk: Predicting International River Basin Conflict and Cooperation. *Global Environmental Politics, 14*(4), 116–138.

Biermann, F., Kim, R. E., Abbott, K. W., Hollway, J., Mitchell, R. B & Scobie, M.. (2020). Taking Stock and Moving Forward. In F. Biermann & R. E. Kim (Eds.), *Architectures of Earth System Governance* (pp. 299–321). Cambridge: Cambridge University Press.

Block, P., Koskinen, J. H., Hollway, J., Steglich, C. E. G., & Stadtfeld, C. (2018). Change We Can Believe in: Comparing Longitudinal Network Models on Consistency, Interpretability and Predictive Power. *Social Networks, 52*(January), 180–191.

Block, P., Stadtfeld, C., & Snijders, T. A. B. (2016). Forms of Dependence: Comparing SAOMs and ERGMs from Basic Principles. *Sociological Methods and Research, 48*(1), 202–239.

Bréthaut, C. (2016). River Management and Stakeholders' Participation: The Case of the Rhone River, a Fragmented Institutional Setting. *Environmental Policy and Governance, 26*(4), 292–305.

Brochmann, M., & Gleditsch, N. P. (2012). Shared Rivers and Conflict—A Reconsideration. *Political Geography, 31*(8), 519–527.

Burt, R. S. (2004). Structural Holes and Good Ideas. *The American Journal of Sociology, 110*(2), 349–399.

Butts, C. T. (2008). A Relational Event Framework for Social Action. *Sociological Methodology, 38*, 155–200.

Cartwright, D., & Harary, F. (1956). Structural Balance A Generalization of Heider's Theory. *Psychological Review, 63*(5), 277–293.

Conca, K. (2005). *Governing Water* (Contentious Transnational Politics and Global Institution Building). Cambridge, MA: MIT Press.

Cornwell, B. (2015). *Social Sequence Analysis* (Methods and Applications). Cambridge: Cambridge University Press.

De Stefano, L., Edwards, P., de Silva, L., & Wolf, A. T. (2010). Tracking Cooperation and Conflict in International Basins: Historic and Recent Trends. *Water Policy, 12*(6), 871–884.

Dellapenna, J., & Gupta, J. (2008). Toward Global Law on Water. *Global Governance, 14*(4), 437–454.

Dinar, S., Dinar, A., & Kurukulasuriya, P. (2011). Scarcity and Cooperation Along International. *Rivers: An Empirical Assessment of Bilateral Treaties1, 55*(3), 809–833.

Fischhendler, I. (2004). Legal and Institutional Adaptation to Climate Uncertainty: A Study of International Rivers. *Water Policy, 6*(4), 281–302.

Fischhendler, I. (2008). When Ambiguity in Treaty Design Becomes Destructive: A Study of Transboundary Water. *Global Environmental Politics, 8*(1), 111–136.

Furlong, K., Gleditsch, N. P., & Hegre, H. (2006). Geographic Opportunity and Neomalthusian Willingness: Boundaries, Shared Rivers, and Conflict. *International Interactions, 32*(1), 79–108.

Gleditsch, N. P., Furlong, K., Hegre, H., Lacina, B., & Owen, T. (2006). Conflicts Over Shared Rivers: Resource Scarcity or Fuzzy Boundaries? *Political Geography, 25*(4), 361–382.

Gould, R. V. (2002). The Origins of Status Hierarchies: A Formal Theory and Empirical Test. *The American Journal of Sociology, 107*(5), 1143–1178.

Granovetter, M. S. (1985). Economic Action and Social Structure: The Problem of Embeddedness. *The American Journal of Sociology, 91*(3), 481–510.

Hansen, H. E., Mitchell, S. M., & Nemeth, S. C. (2008). IO Mediation of Interstate Conflicts. *Journal of Conflict Resolution, 52*(2), 295–325.

Harrigan, N., & Yap, J. (2017). Avoidance in Negative Ties: Inhibiting Closure, Reciprocity, and Homophily. *Social Networks, 48*, 126–141.

Hensel, P. R., & Brochmann, M. (2009). Peaceful Management of International River Claims. *International Negotiation, 14*(2), 393–418.

Hollway, J., & Stadtfeld, C. (2017). Multilevel Network Dynamics and the Evolution of Environmental Cooperation. In *European Workshops in International Studies* (pp. 1–29). Cardiff, UK.

Ingold, K., Fischer, M., de Boer, C., & Mollinga, P. P. (2016). Water Management Across Borders, Scales and Sectors: Recent Developments and Future Challenges in Water Policy Analysis. *Environmental Policy and Governance, 26*(4), 223–228.

Kalbhenn, A. (2011). Liberal Peace and Shared Resources a Fair-Weather Phenomenon? *Journal of Peace Research, 48*(6), 715–735.

Kalbhenn, A., & Bernauer, T. (2012). International Water Cooperation and Conflict: A New Event Dataset. *SSRN Electronic Journal.* https://doi.org/10.2139/ssrn.2176609.

Knieper, C., & Pahl-Wostl, C. (2016). A Comparative Analysis of Water Governance, Water Management, and Environmental Performance in River Basins. *Water Resources Management, 30*(7), 1–17.

Koremenos, B., Lipson, C., & Snidal, D. (2001). The Rational Design of International Institutions. *International Organization, 55*(4), 761–799.

Lubell, M. N. (2013). Governing Institutional Complexity: The Ecology of Games Framework. *Policy Studies Journal, 41*(3), 537–559.

Lubell, M. N., Scholz, J. T., Berardo, R., & Robins, G. L. (2012). Testing Policy Theory with Statistical Models of Networks. *Policy Studies Journal, 40*(3), 351–374.

Lusher, D., Koskinen, J. H., & Robins, G. L. (2013). *Exponential Random Graph Models for Social Networks: Theory, Methods and Applications.* Cambridge: Cambridge University Press.

Mansfield, E. D., Milner, H. V., & Rosendorff, B. P. (2002). Why Democracies Cooperate More: Electoral Control and International Trade Agreements. *International Organization, 56*(3), 477–513.

Mitchell, R. B., & Keilbach, P. M. (2001). Situation Structure and Institutional Design: Reciprocity, Coercion, and Exchange. *International Organization, 55*(4), 891–917.

Munia, H., Guillaume, J. H. A., Mirumachi, N., Porkka, M., Wada, Y., & Kummu, M. (2016). Water Stress in Global Transboundary River Basins: Significance of Upstream Water Use on Downstream Stress. *Environmental Research Letters, 11*(1), 014002–014014.

Rayner, S., Lach, D., & Ingram, H. (2005). Weather Forecasts Are For Wimps: Why Water Resource Managers Do Not Use Climate Forecasts. *Climatic Change, 69*(2), 197–227.

Schmeier, S., Gerlak, A. K., & Blumstein, S. (2015). Clearing the Muddy Waters of Shared Watercourses Governance: Conceptualizing International River Basin Organizations. *International Environmental Agreements: Politics, Law and Economics, 16*(4), 597–619.

Schmeier, S., & Shubber, Z. (2018). Anchoring Water Diplomacy The Legal Nature of International River Basin Organizations. *Journal of Hydrology, 567*(December), 114–120.

Schrodt, P. A. (2012). Precedents, Progress, and Prospects in Political Event Data. *International Interactions, 38*(4), 546–569.

Simmel, G. (1950). *The Sociology of Georg Simmel*. Glencoe: Free Press.

Snijders, T. A. B., Van de Bunt, G. G., & Steglich, C. E. G. (2010). Introduction to Stochastic Actor-Based Models for Network Dynamics. *Social Networks, 32*(1), 44–60.

Soliev, I., Theesfeld, I., Wegerich, K., & Platonov, A. (2017). Dealing with "Baggage" in Riparian Relationship on Water Allocation: A Longitudinal Comparative Study from the Ferghana Valley. *Ecological Economics, 142*(December), 148–162.

Stadtfeld, C., & Block, P. (2017). Interactions, Actors, and Time: Dynamic Network Actor Models for Relational Events. *Sociological Science, 4*, 318–352.

Stadtfeld, C., Hollway, J., & Block, P. (2017a). Dynamic Network Actor Models: Investigating Coordination Ties through Time. *Sociological Methodology, 47*, 1–40.

Stadtfeld, C., Hollway, J., & Block, P. (2017b). Rejoinder: DYNAMs and the Grounds for Actor-oriented Network Event Models. *Sociological Methodology, 47*(1), 56–67.

Tir, J., & Stinnett, D. M. (2012). Weathering Climate Change: Can Institutions Mitigate International Water Conflict? *Journal of Peace Research, 49*(1), 211–225.

Uzzi, B., & Lancaster, R. (2003). Relational Embeddedness and Learning: The Case of Bank Loan Managers and Their Clients. *Management Science, 49*(4), 383–399.

Van Ast, J. A. (1999). Trends Towards Interactive Water Management; Developments in International River Basin Management. *Journal of Hydrology, 24*(6), 597–602.

Verweij, M. (2017). The Remarkable Restoration of the Rhine: Plural Rationalities in Regional Water Politics. *Water International, 42*(2), 207–221.

Wolf, A. T. (1997). International Water Conflict Resolution: Lessons from Comparative Analysis. *International Journal of Water Resources Development, 13*(3), 333–366.

Wolf, A. T. (1998). Conflict and Cooperation Along International Waterways. *Water Policy, 1*(2), 251–265.

Wolf, A. T., Yoffe, S. B., & Giordano, M. (2003). International Waters: Identifying Basins at Risk. *Water Policy, 5*(1), 29–60.

Yoffe, S. B., Wolf, A. T., & Giordano, M. (2003). Conflict and Cooperation Over International Freshwater Resources: Indicators of Basins at Risk. *Journal of the American Water Resources Association, 39*, 1109–1126.

Zawahri, N. A., & Mitchell, S. M. (2011). Fragmented Governance of International. *Rivers: Negotiating Bilateral versus Multilateral Treaties, 55*, 835–858.

Zeitoun, M., & Mirumachi, N. (2008). Transboundary Water Interaction I: Reconsidering Conflict and Cooperation. *International Environmental Agreements: Politics, Law and Economics, 8*(4), 297–316.

Zeitoun, M., & Warner, J. (2006). Hydro-hegemony a Framework for Analysis of Trans-boundary Water Conflicts. *Water Policy, 8*(5), 435–460.

5

Identifying Subsystems and Crucial Actors in Water Governance: Analysis of Bipartite Actor—Issue Networks

Mario Angst and Manuel Fischer

Introduction

Water governance challenges tend to involve more than one policy issue. They are likely to involve important trade-offs between different issues related to a specific challenge. For example, the water and environmental quality in a Western European wetland can depend on nearby farming activities and their use of water, land, and pesticides. It can depend on flood protection measures, which potentially decrease connectivity within the wetland and need land and built infrastructure. Further, it is often also directly related to the management of invasive species, which

M. Angst (✉)
Swiss Federal Institute for Forest, Snow and Landscape Research WSL, Davos Dorf, Switzerland

University of Bern, Bern, Switzerland
e-mail: mario.angst@wsl.ch

© The Author(s) 2020
M. Fischer, K. Ingold (eds.), *Networks in Water Governance*, Palgrave Studies in Water Governance: Policy and Practice, https://doi.org/10.1007/978-3-030-46769-2_5

endanger native species in the wetland, the presence of drinking water sources, and recreational activities such as hiking paths or fishing in the region (Jaramillo et al. 2019).

The above example is representative, but not exhaustive, of the amount of different issues related to the problem of the governance and management of a given wetland ecosystem. The same is true for governance across completely different water-related policy sectors such as urban water management or hydropower provision. Given the many different ecological processes related to governance challenges in the water domain and the many different societal pressures on the resource water, we are likely to observe a high degree of issue-multidimensionality in any given water governance situation.

Water-related problems are certainly not the only policy problems that are concerned with a multitude of interrelated issues. Any policy system can be subdivided into many different issues taken into account to varying degrees by different actors within an interconnected and, usually, multi-level institutional landscape (Lubell 2013). A policy system can thus essentially be conceived of as an almost infinite set of interconnected actors that potentially deal with an almost infinite set of interconnected issues (Angst 2020).

This large network of actors and issues representing an entire policy system or a water governance system is not unstructured. First, there is likely some clustering to be observed due to institutional factors such as borders of political systems or administrative silos (Lafferty and Hovden 2003), due to separate phases of a policy process, or due to incentives for actors to specialize (Howlett et al. 2009). Second, some meaningful delineation of components within a large network of actors and issues is needed if a policy analyst wants to study aspects of this system, such as policy change, coalition formation, or agenda setting dynamics and

M. Fischer
Federal Institute of Aquatic Science and Technology,
Eawag, Dübendorf, Switzerland

University of Bern, Bern, Switzerland
e-mail: manuel.fischer@eawag.ch

related punctuations. Theories of the policy process (e.g., Sabatier and Weible 2014) therefore often define policy subsystems or policy sectors to delineate which issues and which actors belong to a given entity and should consequently be analyzed together to better understand the governance of the issues and related policy outputs. Which issues—and, accordingly, which actors belong to a given subsystem or sector—is a matter of perception and negotiation among relevant actors. Thus, within such a given subsystem, some actors will be especially important as they connect the different issues and actors that belong to this subsystem. Besides identifying subsystems, in this chapter, we therefore identify crucial actors that are "within-subsystem connectors", and can thus enhance coordination and the potential for collective action within a subsystem.

As the above definition of a policy system—a large set of issues and actors—suggests, once an entity like a policy subsystem or a policy sector is "defined" (whether it reflects the political reality and/or serves as a heuristic to the analyst), and relevant actors and issues are assigned to such a substructure, other actors and issues around that substructure should not be forgotten. The surrounding context can influence what happens within a given subsystem, and decisions taken within the subsystem might have implications for politics beyond the focal subsystem. For example, the policy process theory of the Advocacy Coalition Framework (Sabatier and Weible 2014) defines events outside of the subsystem as one of the main potential causes for policy change within the subsystem. Or, to consider another example, diffusion theory emphasizes the importance of what is going on outside of a political system for understanding what decisions are taken inside (e.g., Shipan and Volden 2008; Jones and Jenkins-Smith 2009). Also, from a normative point of view, and mostly related to environmental governance, policy integration (Tosun and Lang 2017)—that is, the coordination of policies from different sectors, as well as the integration of environmental concerns into other policy sectors (Lafferty and Hovden 2003; Jordan and Lenschow 2010)—is claimed to be important for reaching sustainable solutions. Finally, and specifically for the water sector, concepts of Integrated Water Resource Management (Hering and Ingold 2012; Ingold et al. 2016; Lubell and Edelenbos 2013) emphasize the importance of governing and managing different interrelated aspects of water management as a whole. For these reasons, actors that are able to connect different subsystems are as important as actors that can keep subsystems together. We thus also identify and discuss "between-subsystem connectors" that are important for fostering coordination among subsystems, cross-subsystem learning, or policy integration in this chapter.

In this chapter, we demonstrate how large policy systems can be represented as bipartite networks of actors and issues, and how such depiction can be useful for understanding the structure of such systems. We use the case of Swiss water governance to demonstrate how a bipartite network representation of a large governance system can be used to inductively identify (a) subsystems and (b) crucial actors with specific roles within and between these subsystems. Crucial actors are defined as those that are central within a subsystem ("within-subsystem connectors"), and those that are brokers between subsystems ("between-subsystem connectors") (see also, McAllister et al. (2015) for a similar conceptualization). More specifically, we map and analyze Swiss water governance as a network between a set of actors and a set of governance activities of these actors. The empirical study includes 26 issues that are relevant in water politics, such as hydropower, wastewater, energy, or agriculture. It is based on 313 survey responses of organizational actors that indicated their activities with respect to these issues, as well as with respect to the different levels of governance and different phases of the policy process.

In terms of Social Network Analysis, this chapter emphasizes the concept of bipartite networks and the identification of clusters and central nodes in them. Bipartite networks are a type of network consisting of two types of nodes belonging to two non-overlapping sets. Ties can exist between a node from one set and a node of the second set, but not within a set. In our study, nodes in the first set are actors (federal agencies, interest groups, municipalities, cantonal agencies, firms, etc.) involved in water politics at the Swiss national and cantonal levels. Nodes in the second set are water-related issues in Swiss water politics, as identified from a document analysis. Ties between actors and issues are defined as "actors dealing with issues", that is, being involved in the management of an issue, having a stake in the issue, and so on. We first identify subsystems by splitting the network into modules using a bipartite application of Newman's (2006) modularity algorithm. We then explore two simple measures to identify crucial actors originally developed for bipartite networks in ecological studies, such as plant-pollinator or industrial trade networks (Beckett 2016; Guimerà and Nunes Amaral 2005).

The remainder of this chapter is structured as follows. After a theoretical discussion of policy systems and subsystems therein, we discuss the importance of within- and cross-subsystem dynamics. Based on this, we define two types of crucial actors, one that is crucial within subsystems, and one between subsystems. We then discuss the network concept of bipartite networks and its use as a description of governance systems. In the methodological part, we present the Swiss case study as well as our data gathering procedure based on documents and surveys. We then explain how we used this data to construct a bipartite actor—issue network, and how we inductively identify subsystems through bipartite modularity, and crucial actors. We then present our results and discuss the subsystem structure as well as single actors identified as crucial based on our two measures. The final section presents conclusions with respect to water governance, the use of bipartite network data, as well as the practical implications of our findings.

Theory

Subsystems in an Overall Governance System

In order to structure and reduce the complexity of entire policy systems, most theories of the policy process (Sabatier and Weible 2014) focus on single policy sectors, policy subsystems, or policy domains (we use the term "policy subsystem" in the remainder of this chapter). A policy subsystem is the unit of analysis to study advocacy coalition formation or maintenance, policy learning, and change in a policy process (Sabatier and Jenkins-Smith 1993). A policy subsystem spans a geographical area (e.g., local, regional, or national jurisdiction), includes actors involved in the specific policy-making, and is about one specific issue (e.g., migration, water, energy; Jenkins-Smith et al. 2017). For example, an agricultural policy subsystem is composed of farmers' lobbying organizations, the state agency in charge of these issues, as well as various health, development, and other interest groups. We define an issue as an element that specifies the content of a political interaction or negotiation among two or more actors, in contrast to institutions or procedural principles of a policy system that describe the context for these interactions.

Several factors related to the functioning of policy systems and processes lead to substructures within those policy systems. First, actors active in politics usually have incentives and constraints that lead them to specialize in one or a few issues. Actors usually do not have enough resources such as knowledge, personnel, time, or money to engage in more than a few issues (Zhu 1992; see also Henning 2009). Once actors have specialized in a given issue, path dependency creates incentives for actors to continue specializing in that specific issue, as costs of changing issue attention and specialization would be too high. The consequence of actors' being specialized in one or a few issues is that actors cluster around issues within the entire policy system, and thus form subsystems. Geographical borders also lead to substructures, as actors with formal authority or constituencies within one geographical area (e.g., a country or a sub-state within a country) cluster together, whereas other actors belonging to another geographical area form their own cluster, even if both sets of actors deal with the same issue. Finally, and similarly to geographical borders, functional borders within administrative structures lead to subsystems. Given the usually rational-hierarchical organization of public administration and related principles of accountability (Bovens et al. 2014), competencies for given issues are clearly assigned to one or another branch of the public administration system. All of these factors suggest that subsystems are part of the reality of politics.

Dynamics of Subsystem Structures

A subsystem structure is, of course, never set in stone, but evolves over time. This is not least due to actors' interest and related strategies in subsequent policy processes of connecting or separating given issues and subsystems. Some actors might develop roles as "entrepreneurs" benefitting from linking previously unconnected issues (Jones and Jenkins-Smith 2009). For example, some actors might be interested in connecting the subsystem of the pension system and the subsystem of tax policy, because this allows them to propose a policy in line with their interests. Other actors, by contrast, may want to keep the two issues separated. From time to time, actors have to deal with new issues on the political agenda. For example, new problems might emerge on the public and political agendas due to natural developments such as climate change, or to technical developments such as artificial intelligence. The relevant literature claims that some new issues are absorbed by existing subsystems, while others provoke the emergence of new, nascent subsystems (Ingold et al. 2017). An issue is absorbed by an existing subsystem (or even creates a new subsystem) as soon as it is linked to a societal problem that asks for a political solution (i.e., a policy), and imposes a threat to the beliefs of one or more coalitions in the subsystem (see Weible and Ingold 2018). While the temporal evolution of subsystems goes beyond the scope of our analysis, recognizing their dynamic nature highlights the significance of knowing which actors are important within as well as between subsystems at a given point in time.

Finally, the recognition of the dynamic nature of subsystems and their changes over time requires careful consideration of how subsystems should be empirically identified. Evidently, there is no constant structure of subsystems. Still, very often, subsystems are defined in a top-down and approximate way. Top-down because the researcher—based on substantive case knowledge—identifies issues, actors, and processes that in her or his view belong together, instead of examining data. Approximate because a given policy process is often used as a proxy for a subsystem, as a given specific process functions as a visible manifestation of the subsystem to the researcher (e.g., Fischer and Sciarini 2016; Fischer 2015). It remains unclear to what degree a given policy process covers the entire meaningful space of a subsystem. While such top-down and approximate approaches are often the only feasible way to identify subsystems for researchers, this chapter illustrates an alternative, bottom-up approach based on empirically observed organizational activity. Subsystem identification is thus approached as an empirical question. In cases when the necessary data is available, this represents a more appropriate and precise way of identifying subsystems within entire policy systems (Angst 2020).

The Importance of Cross-subsystem Dynamics

In the literature, cross-subsystem dynamics appear under different labels such as cross-sector interactions and overlapping subsystems. For example, Zafonte and Sabatier (1998) argue that "overlapping and nested subsystems" make actors from different subsystems mutually interdependent and incentivize coordination among them. Similarly, "trans-subsystem dynamics" (Jones and Jenkins-Smith 2009) in terms of feedback and spillover effects across subsystems are claimed to influence policy change. Also, logics of "sector intersection" are argued to be important for understanding how the structure of coalitions in one sector influences coalition structures in other sectors (Hoberg and Morawski 1997). Finally, the "ecology of games" approach (Dutton et al. 2012; Lubell 2013; Berardo and Lubell 2019) posits that actors deliberately choose to participate in parallel games according to their specific interests and make incremental decisions about many interrelated topics (Dutton et al. 2012). These approaches emphasize the importance of a precise bottom-up identification of subsystems. It furthermore points to the crucial role of actors that are able to connect different subsystems.

Two Types of Crucial Actors

The identification of crucial actors within governance networks rests on the assumption that not all actors' positions within a network are equal. Some actors occupy structural positions within networks that allow them to influence outcomes and processes of natural governance more than others (Bodin and Crona 2009). Such structural power can be linked to the concept of social capital, which is individually realized in the network positions of actors (Burt 2000). Accordingly, analysts often investigate crucial actors in two dimensions: bonding and bridging (Berardo 2014; Scott and Thomas 2017).

On the one hand, central actors with many bonding ties are highly connected within dense social structures. In network analysis terms, this often refers to actors with high degree centrality (a high number of ties) in structures with high transitive closure (closed triangles, e.g., actors in a group generally have contacts with all other group actors). Such actors with a large number of bonding ties face low transaction costs in their actions, as they can leverage connections of trust and reciprocity.

On the other hand, actors with many bridging ties have contacts in different parts of the network—they serve as bridges. They are often discussed in the literature in the context of brokerage concepts. Brokerage in networks generally describes actors that connect otherwise disconnected parts of networks (Gould and Fernandez 1989; Everett and Valente 2016). In the policy literature, brokers are mostly referred to as creating linkages between opposing coalitions of actors, and thus enabling negotiations or compromise finding in a policy process (Ingold and Varone 2011; Christopoulos and Ingold 2015). Betweenness centrality is often used as a simple measure to identify such brokers, as the measure assesses how often an actor is positioned on the shortest path between any other actors in the network (Freeman 1978). However, brokerage (or bridging) is a multi-dimensional concept, and measures that are more precise about the exact nature of brokerage have thus been developed for social networks in general (Gould and Fernandez 1989; Everett and Valente 2016), for applications in bipartite networks (Jasny and Lubell 2015), or specifically for governance networks (McAllister et al. 2015; Angst et al. 2018). In policy networks, brokers are often actors on higher levels of governance, as exemplified by studies of land use planning in Swiss mountain regions (Ingold 2014) and park projects (Hirschi 2010). Also, governmental organizations often possess the necessary manpower and resources to undertake coordinative activities, as that is what many of them are tasked with (Ingold 2011).

Bonding and bridging are not mutually exclusive, and are often related—in fact, a classic argument is that often only bridging ties allow for the greatest individual exploitation of bonding social capital for an actor (Burt 2000). Building on this literature, we conceptualize two different kinds of categories of crucial actors from a subsystem perspective: within- and between-subsystem connectors. Within-subsystem connectors fit the bonding narrative more closely. They organize an individual subsystem, and are highly connected within it. From our bipartite perspective on policy systems, consisting of actors and issues, such actors are involved in many issues within a subsystem, but not beyond it. These organizations thus often develop specialized forms of knowledge related to their subsystem, and are important actors in shaping processes within a subsystem.

Between-subsystem connectors, our second category, fit into the bridging dimensions. They connect subsystems—in our case by being involved in issues in multiple subsystems. Such actors appear, for example, as boundary-penetrating organizations in Jones and Jenkins-Smith (2009). These actors have, due to their involvement in a more diverse set of issues, the potential to transfer knowledge between different subsystems, but also to exploit linked action situations (Kimmich 2013) in different subsystems to their advantage. Again, actors can also be both within- and between-subsystem connectors—and such actors function as a type of a super-connector, and are likely to yield considerable influence across the entire policy system.

Case, Data, and Methods

Water Governance in General

This chapter does not focus on a specific issue of water governance, but rather aims to cover the entire field of water governance—that is, all issues that are somehow related to water. These issues include the usual water-related issues such as flood protection, urban water management, and water protection, but also issues that do not solely pertain to water, but in which water does play a role, such as energy, agriculture, biodiversity, or health. We thus avoid pre-defining a single policy subsystem dealing with a given aspect of water governance in a top-down way. Rather, our goal is to identify relevant subsystems as well as crucial actors within the water governance system in a bottom-up way.

The Case of Swiss Water Governance

Switzerland is the "Water Castle" of Europe, as several main watercourses, such as the Rhone or Rhine rivers, have their origin in the Swiss Alps.

Through its integrative and consensus-oriented direct-democratic system and federalist structure, the country has the institutional pre-conditions to account for the multi-level and boundary spanning nature of water (Kriesi and Trechsel 2008; Sciarini et al. 2015). Switzerland is a small country where water scarcity or floods also have an immediate impact on issues such as water pollution or protection. When it comes to water governance, natural resources management, and environmental politics, the Swiss national and cantonal levels share a complex set of competences (Linder and Vatter 2001; Fischer et al. 2010). Water-related competences at the national level have increased since the end of the nineteenth century when the federal government started to expect them from the cantons. Over time, the federal government has formulated general principles on flood protection, fisheries, hydropower, and water-related land use and planning. Issues of water protection and quality are more recent competences at the level of the federal government (Mauch and Reynard 2004). Cantons still are responsible for the implementation of federal laws, and often benefit from high flexibility and financial compensations for their tasks. Cantons also remain the formal owners of most water bodies (Mauch and Reynard 2004). Due to the fact that water politics in Switzerland takes place on many different levels, and that actors perform different activities depending on the level they are at, our analysis takes into account different levels of governance as well as different phases of the policy process.

Data Gathering: Swiss Water Governance Issues and Actor Survey

We gathered a complete set of water-relevant issues in Switzerland through a bottom-up document analysis procedure (see Brandenberger et al. (2020) for a detailed description). A team of coders analyzed newspaper articles and parliamentary protocols related to three keywords: water, lake, and water body. The keyword search covered the national parliament and a leading newspaper covering the whole country, as well as the cantonal parliament of Bern and a cantonal newspaper in Bern. This ensured that aspects of water governance on all federal levels were covered. The coding procedure resulted in a list of 26 issues and 56 sub-issues covering an encompassing variety of aspects of Swiss water governance, ranging from recreational boating to hydropower plant construction (Table 5.1 presents the list of issues). At the same time, the coders also marked the names of any organizational actor occurring in relation to an issue in any document. This provided us with a starting sample of organizational actors involved in Swiss water governance.

We used the information gathered in the document analysis to conduct a nation-wide online survey among actors in water governance. We surveyed organizations including administrative agencies on various jurisdictional levels, municipalities, civil society organizations, service providers, and engineering firms. After the first round of surveys in the summer of 2016, we used snowball sampling to identify previously unidentified actors (the survey asked actors to name other relevant

Table 5.1 List of issues in Swiss water governance considered in this paper, derived from document analysis

Issue
Aquatic habitat protection
Artificial snow
Biodiversity impacts agriculture
Energy politics
Environmental laws impact quality
Fish biodiversity
Flood protection implementation
Fracking
Glacier retreat
Hydropower construction
Hydropower impacts
Hydropower profitability
Landscape protection
Micropollution
Protection from pollution
Renaturation for flood protection
Snow clearance
Spatial planning floods
Subterranean resources
Touristic water use
Water supply infrastructure
Water supply planning
Water supply reorganization
Water trade
Water use reduction

organizations in their field they had either exchanged information with or regarded as allies or opponents in issues they were active in). We then conducted a second survey round in the spring of 2017. Over both rounds, we sent the survey to 476 organizations, of which 313 participated. The response rates for the initial and snowballing rounds were 69% and 64%, respectively. The survey asked organizations to indicate issues and sub-issues in water governance they had regularly encountered in water projects carried out in the three years preceding 2016. For every issue they chose among the 26 issues gathered in the document analysis, the survey asked organizations to indicate on which of these levels they normally dealt with each issue (municipal, cross-municipal, state, cross-state, or national). Similarly, the survey asked organizations to indicate the phases where they had normally engaged with each issue (initiation, planning, decision-making, implementation, or evaluation).

Bipartite Actor—Issue Networks

Our data gathering resulted in detailed data about actor activity. We obtained data about the issues that each actor is engaged in. Furthermore, for every one of these issues, we know on which level and at which phase of the policy process the actor is generally active. We used this information to distinguish all uniquely occurring triplet combinations of issues, levels, and phases (e.g., flood protection implementation on the municipal level or biodiversity protection evaluation on the national level). Substantively, this means that issues are subdivided to even more detailed and precise elements, at least for analytical purposes. That is, there is a flood protection issue on the national level and in the implementation phase, which is different from a flood protection issue on the national level and in the decision-making phase. We then created a rectangular incidence matrix with actors (rows) and all unique such triplets, indicating every actor's detailed activity profiles and drawing up a bipartite graph between actors and triplets. The level of detail allowed by this representation is essential in dealing with policy systems characterized by multiple levels and actors working at different phases of the policy process. Sometimes, differentiation in a governance system is as stark between levels within the same issue as it is between issues (Angst 2020).

Inductively Identifying Subsystems Through Bipartite Modularity

An interesting aspect of network methodology is that different research communities across disciplines are working on network methods, developing them to suit their needs. However, the language of networks and graph theory is in and of itself neutral, facilitating exchange between very different fields of study. One core field of study across research fields is community detection. In communities, generally, members share many links within a community and comparatively few with members of other communities (Fortunato 2010). For bipartite networks, specialized community detection methods have been developed especially by physicists and also by ecologists (Beckett 2016). The latter because bipartite networks are ubiquitous in the study of ecological networks. Plant-pollinator networks are a prime example. Pollinators (e.g., different species of insects) form one set of nodes and plant species form a second one. The two set of nodes are connected through pollinating activity, leading to a bipartite network. To get a grasp on an ecological system composed of many different species, ecologists thus use bipartite community detection to find subgroups of plants and pollinators that are relatively exclusive parts of the ecological system in terms of pollination. This means that pollinators within a community generally pollinate plants within that community, instead of plants of other communities, and the same applies for plants within a community, which are more often pollinated by community members than by pollinators outside the community.

Based on the same thinking, formalizing a political system as a bipartite network, as we have done for the case of Swiss water governance, makes it possible to apply community detection methods to identify subsystems. Identifying subsystems in an actor-issue network is, from a purely technical standpoint, similar to identifying communities within a plant-pollinator network, if care is taken in the interpretation of the results. Community detection in this regard means finding subgroups of actors and issues where actors generally work on the issues within that community, rather than issues in other communities. These communities of issues and actors correspond to inductively identified subsystems. We identified communities by maximizing Newman's modularity measure as implemented in the R package "bipartite" (Dormann et al. 2008). Modularity is a concept describing how fragmented or cohesive a network is overall. Modularity increases if interactions within subgroups occur more frequently than in a null model. It becomes negative if the opposite is the case. Modularity maximization works by finding a way in which to partition a network such that modularity is maximized. This means that a network is partitioned such that, compared to a null model, no other partition is possible that increases interactions within subgroups.

Clustering approaches to data analysis, including network community detection, always contain some subjective judgment. They yield differing results based on the choice of algorithm and parameters, and there is no single correct solution purely based on the data. However, we would argue that this is also one of the strengths of such approaches—to make sense of bottom-up approaches to subsystem identification, the advantages, and disadvantages of a chosen approach need to be considered based on theory. Different approaches can highlight different aspects of a policy system. The modularity maximization procedure used in this chapter, for example, has advantages in being able to identify subsystems of vastly varying size.

Crucial Actors: Exploring a Measure Developed in Ecology

We identify two different types of crucial actors from a subsystem perspective as outlined above. These are within-subsystem coordinators and between-subsystem coordinators. To do this empirically, we explore a measure developed for ecological networks. Again, the core thinking behind the identification of crucial nodes in a network is analogous to the logic of ecological networks. Returning to the case of pollinator-plant networks, an ecologist might be interested in the pollinators that are between-communities connectors, holding an ecosystem as a whole together, or those that are especially active within their community. Olesen et al. (2007) use two degree-based measures to do exactly that, based on Guimerà and Nunes Amaral (2005), although for the case of unipartite networks. Every network node is assigned a z-score measuring within-community connections and a c-score measuring between-community connections. Both measures are essentially based on standardized counts of within- and between-module degree per node.

The measures were not originally developed for bipartite networks, but can be reasonably well interpreted in our policy subsystems case. As we are only interested in actors (i.e., crucial actors), we only calculate scores for them and thus evaluate only one of the two sets of nodes present in the bipartite networks (as implemented in the bipartite package). Z- and c-scores have direct interpretations in this case—they measure the standardized amounts of ties to policy issues within and outside a subsystem for each actor.

Results and Discussion

Subsystem Identification

Figure 5.1 visualizes the entire bipartite actor-issue network and the location of subsystems within. The visualization is not particularly useful by itself, as is often the case with "hairball" network visualizations. Still, it illustrates the extent of the network and the fact that the modularity optimization algorithm identified five relatively distinct subsystems. We colored links if they occur within subsystems in the color of the subsystem, and in gray if they occurred between actors and issues from different subsystems.

Figure 5.2 plots the content of the subsystems in terms of issues and their characteristics. It shows that the identified subsystems have distinct characteristics that make substantive sense with regard to Swiss water governance. Subsystem 4 is a distinct small subsystem concerned mostly with local-level touristic water use. A closer look at organizations involved in the subsystem shows that it contains mostly local tourism organizations, water sport organizations, and some private operators of activities such as boating excursions. Subsystem 2 is a higher-level subsystem that is mostly concerned with the evaluation of water quality. It contains a number of scientific laboratories, cantonal agencies tasked with water quality control, and fish protection organizations.

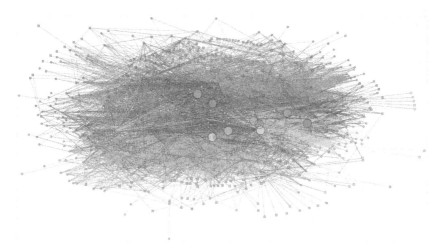

Fig. 5.1 Bipartite actor-issue network. (Note: Bipartite actor-issue network, colored by subsystems, actors with top 1% c-scores (between module connectors) emphasized. Circles indicate actors; squares each indicate an issue at a specific level and phase, or unique triplet combination of issue, level, and phase. For example, this might mean biodiversity protection (issue) evaluation (phase) at the national level (level). Ties between issues and actors indicate that an actor is regularly active in the given issue, during the specific phase, at the specific level)

Fig. 5.2 Composition of five subsystems in Swiss water governance. (Note: Composition of five subsystems in Swiss water governance in terms of characteristics of issues within the subsystem. Subsystems based on modularity maximization in bipartite actor-issue network. Non-white colors indicate the occurrence of issues, levels, and phases in a subsystem. For issues, the strength of colors indicates the relative frequency of a given issue in the subsystem. For level and phases, the strength of colors indicates the relative frequency of a given level or phase for issues present in the subsystem)

Subsystem 3 is predominantly local and deals with water supply, an area where municipal competences are famously strong in Switzerland. It is therefore also not surprising that it contains a large proportion of municipalities and wastewater treatment plants. Subsystem 5 deals with flood protection issues, which happens mostly on the inter-municipal level. It contains a large number of engineering firms and local flood protection organizations ("Schwellengemeinschaften"). Subsystem 1 is a hydropower and energy subsystem. It involves activity on all levels, which again makes sense due to both the national level regulation but also the strong local impacts of hydropower projects. A diverse group of actors from both energy and agricultural sectors, as well as nature and landscape protection interests, is associated with this subsystem.

Crucial Actors—Within- and Between-Subsystem Connectors

C-scores, measuring the amount of within-degree connecting, and z-scores, measuring the amount of between-subsystem connecting, draw up a two-dimensional space in which each actor can be located (see Fig. 5.3). In this figure, we have introduced threshold lines for simple reasons of readability in order to be able to distinguish actors with high values on either of the dimensions from all other actors.[1] We are interested in three types of actors within this space. Pure within-subsystem connectors (what Olesen et al. (2007) call hubs) score high in z-scores. Pure between-subsystem-connectors score high in c-scores. Super-connectors both have high c- and z-scores, meaning that they both connect to a wide variety of issues within their subsystem, as well as to disproportionally many issues in other subsystems.

---

[1] Olesen et al. (2007) suggest z-values of above 2.5 and c-values of above 0.62 as cutoff criterions to classify nodes into these categories. These values are marked in Fig. 5.3 to give a rough overview, but their applicability to bipartite networks, as well as their statistical foundation in general, are questionable.

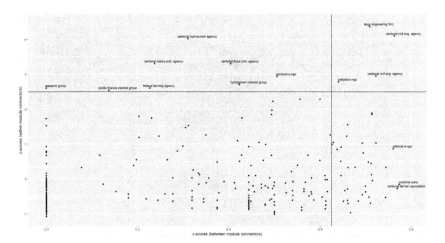

Fig. 5.3 c- and z-score distributions. (Note: c- and z-score distribution for actors in Swiss water governance. Lines at c = 0.625 and z = 2.5 show (non-authoritative) threshold values for crucial actors suggested in Olesen et al. (2007). Indicative labels are shown for actors with z > 2.5 and (to avoid overplotting) c > 0.75)

In the following, we highlight three exemplary actors at the extreme ends of the spectrum. Figure 5.4 (left-hand side) shows the ego network of order two for a pure within-subsystem connector, thus an actor with a high z- but very low c-score. An ego-network of order two only shows a subset of the network containing a focal node (in this case the within-subsystem connector) and all other nodes that are directly or indirectly (two ties at maximum) related to this actor. In our bipartite network case, this includes all issues (shown as squares) the focal actor is involved in, plus all other actors (shown as circles) involved in these issues. Nodes in the figure are colored based on the subsystem to which they belong. The within-subsystem connector shown is a research group working in the field of environmental toxicity. The group is very active in a large number of issues within its subsystem (subsystem 2), playing an important knowledge provision and broker role. However, it does not appear to see its role as extending beyond the subsystem, and does not work on issues outside it.

Figure 5.4 (center) shows the same ego network for a relatively pure between-subsystem connector. As such, the actor is not involved in an extremely large share of issues in the subsystem to which it belongs;

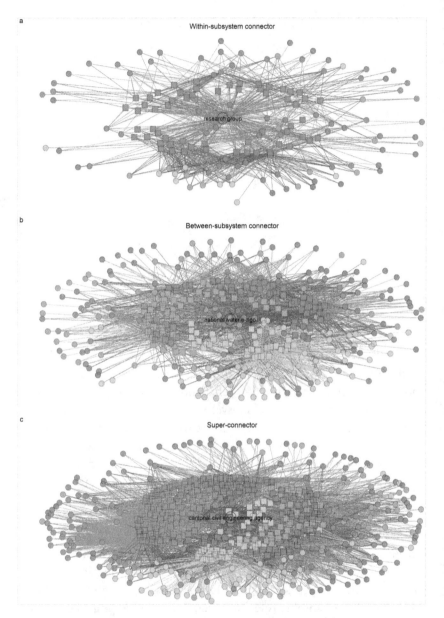

Fig. 5.4 Ego networks within the bipartite actor-issue network (see Fig. 5.1 for details) of Swiss water governance for three exemplary actors. (a) Within-subsystem connector example: research group working on environmental toxicity. (b) Between-subsystem connector example: national level, water-focused environmental NGO. (c) Super-connector example: civil engineering agency within a large Swiss canton

compared to the rest of actors in the subsystem it is a rather peripheral actor. However, the actor is active in a large number of issues from different subsystems, as compared to actors in the network in general. The organization in question here is the largest purely water-focused environmental non-governmental organization (E-NGO) in Switzerland. As environmental and nature protection issues in water governance touch a diverse set of other issues such as hydropower provision, floodwater prevention, or agricultural pollution—as claimed at the outset in terms of the extreme multi-dimensionality of water-related policy problems, it makes sense for this organization to have a broad activity profile in the different subsystems dealing with these issues. Also, due to resource constraints usually present for E-NGOs in terms of personnel and time, it also makes sense that the organization is not involved in every single issue within its focal subsystem (subsystem 1, focused on hydropower and energy). The categorization of this type of boundary-penetrating organization shows the limits of assigning actors into only a single subsystem. It could reasonably be argued that the particular E-NGO is part of at least three subsystems, including flood protection and water quality, in addition to energy and hydropower. The calculation and interpretation of z-scores alleviates this problem to some extent, as it highlights such boundary-penetrating organizations.

Finally, Fig. 5.4 (right-hand side) shows a super-connector with both high z- and c-scores. The actor in question is a cantonal agency in a large canton responsible for civil engineering questions. This result is interesting, as it highlights an actor involved in a very large share of issues both in its main subsystem related to flood protection, but also in many issues in all other subsystems, except for tourism. Still, the agency in question is not even focused primarily on water, but rather deals with various infrastructure projects in general. This case therefore powerfully illustrates two points. First, actors connecting subsystems can sometimes be found in unexpected places, which can be important to consider for practitioners working in multi-dimensional water governance problems, as well as for researchers studying it. Second, it highlights the crucial role that questions of water governance play in infrastructure projects, such as effects on groundwater, flood protection risk, or the loss of aquatic habitats due to building activity. It also shows why both interdisciplinarity and early involvement of a broad base of stakeholder are often highlighted in large infrastructure projects. The fact that the agency also is a within-subsystem connector within its focal subsystem further shows that it also possesses a large amount of focused expertise within a main area, which it ports to other subsystems.

The literature on crucial actors (and similar concepts) emphasizes the importance of the resources (e.g., personnel, time, finances) needed for actors to be able to play either within-subsystem or between-subsystem connector roles. Because they tend to have more resources, studies thus often expect actors on higher levels of governance systems or governmental actors to play these roles (Hirschi 2010; Ingold 2011, 2014; Angst et al. 2018). An example for a between-subsystem connector was an E-NGO active at the national level, and another national level E-NGO is among the four super-connectors. Besides the illustrative example of the super-connector of a cantonal level agency responsible for civil engineering, the same agency from another canton also plays a super-connector role.

Not only national level or governmental actors play crucial roles. A private engineering firm is the fourth super-connector, and an applied research center is the illustrative example of a within-subsystem connector. The crucial role of the private engineering firm might be due to the historically dominant role of engineering in Swiss water politics. Scientific institutions and scientists have been reported to play coordinating roles in subsystems, and even connect different opposing coalitions, especially in low-conflict contexts (Weible et al. 2010; Ingold and Gschwend 2014), which is a fair way to characterize many dimensions of Swiss water politics (Angst 2020).

Conclusions

This chapter starts with the assumption that it is meaningful to conceive of a policy system as a set of actors and issues, with actors being connected to issues as they are somehow "dealing with them". Given the complexity of water governance with its many aspects related to water quality, water use, water protection, and so on, and its manifold interactions with related areas such as biodiversity, energy production, agriculture, or infrastructure, a representation of the sector as a bipartite network of actors and issues seems to make sense. However, we do not claim that water governance is the only governance field characterized by these complexities: in reality, once a researcher digs deeply into a field, many different governance fields, from financial politics to security politics and infrastructure politics, are probably similar. In that respect, we think that our approach as presented in this chapter could be useful for studying any governance field in its entire complexity.

In this chapter, based on the bipartite network, we identify (a) subsystems within the complex policy system around Swiss water governance, and (b) two types of crucial actors that connect actors and issues either within subsystems ("within-subsystem connectors") or between subsystems ("between-subsystem connectors"). We identify these actor based on methods borrowed from ecological studies. Both of these types of actors are important in order to understand the functioning and dynamics of complex policy systems. We thus discuss examples of which actors do actually play such roles. Additionally, we discuss an actor that is a "super-connector", that is, an actor that is both a within- and a between-subsystem connector. The presence of actors that function in this way also implies that both roles are not mutually exclusive—from neither a theoretical nor a methodological point of view. As we have discussed when presenting empirical examples of these crucial actors from Swiss water governance, playing a connector role most likely relies on actors having large amounts of organizational or knowledge resources. We have thus discussed that actors who are between-subsystem connectors are most likely rather peripheral in the specific subsystems they connect. However, we see that at least four actors in our empirical example are able to play both roles, and thus occupy an even more important role in Swiss water governance.

Whether these crucial actors can translate their structurally specific position into real influence in Swiss water governance goes beyond the scope of our analysis. To ascertain that, qualitative case studies of decision-making processes related to the relevant subsystems would be needed, or, alternatively or in combination with the former, a quantitative analysis of actors' influence based, for example, on reputational power (Fischer and

Sciarini 2015). There are many other questions that the analysis in this chapter only points to superficially, and that should be studied in future research, or that are simply not considered by our analysis. Besides power and influence, another important concept in studies on governance and politics is institutions. Institutions such as constitutional rules, formal organizational structures of public administration, or forums where actors gather to discuss relevant issues are not taken into account in this chapter, but could be important for structuring the network of actors and issues (Fischer et al. 2019) and for understanding outcomes of policy systems. Furthermore, our bipartite network takes into account only interactions between actors and issues, and disregards potentially important information on interactions among actors or dependencies among issues (e.g., Bodin 2017). An advantage of our simple actor-issue network is that data gathering for such a model is simpler than collecting all information for a multi-level network of actors and issues that can also be connected among themselves. A final methodological limitation of the subsystem identification approach based on bipartite community detection via modularity maximization is that each actor belongs to a single subsystem only. Given the prominent role of concepts such as policy brokerage (Christopoulos and Ingold 2015) and boundary penetration (Jones and Jenkins-Smith 2009), where actors take part in multiple subsystems, this is likely to be an inadequate description of reality. An alternative approach to subsystem identification, which makes it possible to detect multi-subsystem participation based on issue clustering, using the same data as in this chapter, is described in Angst (2020), and shows that boundary penetration does indeed occur frequently.

Overall, this chapter suggests that policy systems can meaningfully be represented as large networks comprised of actors and issues. They can then be subdivided into subsystems based on the empirically observed network structure, and these subsystems inform us about what water-related issues are dealt with together, and by which actors. Based on this, crucial actors, such as within-subsystem and between-subsystem connectors, can be identified. These are important for the elaboration of policy solutions, and the dynamic evolution of the entire complex water policy system.

References

Angst, M. (2020). Bottom-Up Identification of Subsystems in Complex Governance Systems. *Policy Studies Journal*. Online First (see https://onlinelibrary.wiley.com/doi/full/10.1111/psj.12301).

Angst, M., Widmer, A., Fischer, M., & Ingold, K. (2018). Connectors and Coordinators in Natural Resource Governance: Insights from Swiss Water Supply. *Ecology and Society, 23*(2), 1.

Beckett, S. J. (2016). Improved Community Detection in Weighted Bipartite Networks. *Royal Society Open Science, 3*, 140536.

Berardo, R. (2014). Bridging and Bonding Capital in Two-Mode Collaboration Networks. *Policy Studies Journal, 42*(2), 197–225.

Berardo, R., & Lubell, M. (2019). The Ecology of Games as a Theory of Polycentricity: Recent Advances and Future Challenges. *Policy Studies Journal, 47*(1), 6–26.

Bodin, Ö. (2017). Collaborative Environmental Governance: Achieving Collective Action in Social-ecological Systems. *Science, 357*(6352), eaan1114.

Bodin, Ö., & Crona, B. I. (2009). The role of social networks in natural resource governance: What relational patterns make a difference? *Global Environmental Change, 19*(3), 366–374, ISSN 0959-3780, https://doi.org/10.1016/j.gloenvcha.2009.05.002. http://www.sciencedirect.com/science/article/pii/S0959378009000405

Bovens, M., Goodin, R. E., & Schillemans, T. (Eds.). (2014). *The Oxford Handbook of Public Accountability*. Oxford, UK: Oxford University Press.

Brandenberger, L., Ingold, K., Fischer, M. Schläpfer, I., & Leifeld, P. (2020). Overlapping Network Structures: Why Actors Engage in Diverse Policy Issues. *Policy Studies Journal*. Submitted.

Burt, R. S. (2000). The Network Structure of Social Capital. *Research in Organizational Behavior, 22*(0), 345–423.

Christopoulos, D., & Ingold, K. (2015). Exceptional or Just Well Connected? Political Entrepreneurs and Brokers in Policy Making. *European Political Science Review, 7*(3), 475–498.

Dormann, C. F., Gruber, B., & Fruend, J. (2008). Introducing the Bipartite Package: Analysing Ecological Networks. *R News, 8*(2), 8–11.

Dutton, W. H., Schneider, V., & Vedel, T. (2012). Ecologies of Games Shaping Large Technical Systems: Cases from Telecommunications to the Internet. In J. Bauer, A. Lang, & V. Schneider (Eds.), *Innovation Policy and Governance in High-tech Industries* (pp. 49–75). Heidelberg, Germany: Springer.

Everett, M. G., & Valente, T. W. (2016). Bridging, Brokerage and Betweenness. *Social Networks, 44*, 202–208.

Fischer, M. (2015). Institutions and Coalitions in Policy Processes: A Cross-sectoral Comparison. *Journal of Public Policy, 35*(2), 245–268.

Fischer, M., Angst, M., & Maag, S. (2019). Co-participation in the Swiss Water Forum Network. *International Journal of Water Resources Development, 35*(3), 446–464.

Fischer, M., & Sciarini, P. (2015). Unpacking Reputational Power: Intended and Unintended Determinants of the Assessment of Actors' Power. *Social Networks, 42*, 60–71.

Fischer, M., & Sciarini, P. (2016). Drivers of Collaboration in Political Decision Making: A Cross-sector Perspective. *The Journal of Politics, 78*(1), 63–74.

Fischer, M., Sciarini, P., & Traber, D. (2010). The Silent Reform of Swiss Federalism: The New Constitutional Articles on Education. *Swiss Political Science Review, 16*(4), 747–771.

Fortunato, S. (2010). Community Detection in Graphs. *Physics Reports, 486*(3–5), 75–174.

Freeman, L. C. (1978). Centrality in Social Networks Conceptual Clarification. *Social Networks, 1*(3), 215–239.

Gould, R. V., & Fernandez, R. M. (1989). Structures of Mediation: A Formal Approach to Brokerage in Transaction Networks. *Sociological Methodology, 19*, 89–126.

Guimerà, R., & Nunes Amaral, L. (2005). Functional Cartography of Complex Metabolic Networks. *Nature, 433*, 895–900, https://doi.org/10.1038/nature03288.

Henning, C. (2009). Networks of Power in the CAP System of the EU-15 and EU-27. *Journal of Public Policy, 29*(2), 153–177.

Hering, J. G., & Ingold, K. (2012). Water Resources Management: What Should Be Integrated? *Science, 336*(6086), 1234–1235.

Hirschi, C. (2010). Strengthening Regional Cohesion: Collaborative Networks and Sustainable Development in Swiss Rural Areas. *Ecology and Society, 15*(4), 16.

Hoberg, G., & Morawski, E. (1997). Policy Change Through Sector Intersection: Forest and Aboriginal Policy in Clayoquot Sound. *Canadian Public Administration, 40*(3), 387–414.

Howlett, M., Ramesh, M., & Perl, A. (2009). *Studying Public Policy: Policy Cycles and Policy Subsystems* (Vol. 3). Oxford, UK: Oxford University Press.

Ingold, K. (2011). Network Structures Within Policy Processes: Coalitions, Power, and Brokerage in Swiss Climate Policy. *Policy Studies Journal, 39*(3), 435–459.

Ingold, K. (2014). How Involved are They Really? A Comparative Network Analysis of the Institutional Drivers of Local Actor Inclusion. *Land Use Policy, 39*, 376–387.

Ingold, K., Fischer, M., & Cairney, P. (2017). Drivers for Policy Agreement in Nascent Subsystems: An Application of the Advocacy Coalition Framework to Fracking Policy in Switzerland and the UK. *Policy Studies Journal, 45*(3), 442–463.

Ingold, K., Fischer, M., de Boer, C., & Mollinga, P. P. (2016). Water Management Across Borders, Scales and Sectors: Recent Developments and Future Challenges in Water Policy Analysis. *Environmental Policy and Governance, 26*(4), 223–228.

Ingold, K., & Gschwend, M. (2014). Science in Policy-making: Neutral Experts or Strategic Policy-makers? *West European Politics, 37*(5), 993–1018.

Ingold, K., & Varone, F. (2011). Treating Policy Brokers Seriously: Evidence from the Climate Policy. *Journal of Public Administration Research and Theory, 22*(2), 319–346.

Jaramillo, F., Desormeaux, A., Hedlund, J., Jawitz, J. W., Clerici, N., Piemontese, L., Celi, J., et al. (2019). Priorities and Interactions of Sustainable Development Goals (SDGs) with Focus on Wetlands. *Water, 11*(3), 619.

Jasny, L., & Lubell, M. (2015). Two-Mode Brokerage in Policy Networks. *Social Networks, 41*, 36–41.

Jenkins-Smith, H., Weible, C. M., Nohrstedt, D., & Ingold, K. (2017). The Advocacy Coalition Framework—An Overview of the Research Program. In P. A. Sabatier & C. M. Weible (Eds.), *Theories of the Policy Process* (4th ed., pp. 135–172). New York, NY: Routledge.

Jones, M. D., & Jenkins-Smith, H. C. (2009). Trans-subsystem Dynamics: Policy Topography, Mass Opinion, and Policy Change. *Policy Studies Journal, 37*(1), 37–58.

Jordan, A., & Lenschow, A. (2010). Environmental Policy Integration: A State of the Art Review. *Environmental Policy and Governance, 20*(3), 147–158.

Kimmich, C. (2013). Linking Action Situations: Coordination, Conflicts, and Evolution in Electricity Provision for Irrigation in Andhra Pradesh, India. *Ecological Economics, 90*, 150–158.

Kriesi, H., & Trechsel, A. H. (2008). *The Politics of Switzerland: Continuity and Change in a Consensus Democracy*. Cambridge: Cambridge University Press.

Lafferty, W., & Hovden, E. (2003). Environmental Policy Integration: Towards an Analytical Framework. *Environmental Politics, 12*(3), 1–22.

Linder, W., & Vatter, A. (2001). Institutions and Outcomes of Swiss Federalism: The Role of the Cantons in Swiss Politics. *West European Politics, 24*(2), 95–122.

Lubell, M. (2013). Governing Institutional Complexity: The Ecology of Games Framework. *Policy Studies Journal, 41*(3), 537–559.

Lubell, M., & Edelenbos, J. (2013). Integrated Water Resources Management: A Comparative Laboratory for Water Governance. *International Journal of Water Governance, 1*(3–4), 177–196.

Mauch, C., & Reynard, E. (2004). The Evolution of the Water Regime in Switzerland. In I. Kissling-Näf & S. Kuks (Eds.), *The Evolution of National Water Regimes in Europe. Environment and Policy* (Vol. 40, pp. 293–328). Dordrecht, The Netherlands: Springer.

McAllister, R. R., Taylor, B. M., & Harman, B. P. (2015). Partnership Networks for Urban Development: How Structure is Shaped by Risk. *Policy Studies Journal, 43*(3), 379–398.

Newman, M. E. J. (2006). Modularity and Community Structure in Networks. *Proceedings of the National Academy of Sciences, 103*(23), 8577–8582.

Olesen, J. M., Bascompte, J., Dupont, Y. L., & Jordano, P. (2007). The Modularity of Pollination Networks. *Proceedings of the National Academy of Sciences, 104*(50), 19891–19896.

Sabatier, P. A., & Jenkins-Smith, H. C. (1993). *Policy Change and Learning: An Advocacy Coalition Approach*. Boulder, CO: Westview Press.

Sabatier, P. A., & Weible, C. M. (Eds.). (2014). *Theories of the Policy Process*. Boulder, CO: Westview Press.

Sciarini, P., Fischer, M., & Traber, D. (2015). *Political Decision-making in Switzerland: The Consensus Model under Pressure*. Springer Nature EN and Palgrave Macmillan.

Scott, T. A., & Thomas, C. W. (2017). Winners and Losers in the Ecology of Games: Network Position, Connectivity, and the Benefits of Collaborative Governance Regimes. *Journal of Public Administration Research and Theory, 27*(4), 647–660.

Shipan, C. R., & Volden, C. (2008). The Mechanisms of Policy Diffusion. *American Journal of Political Science, 52*(4), 840–857.

Tosun, J., & Lang, A. (2017). Policy Integration: Mapping the Different Concepts. *Policy Studies, 38*(6), 553–570, https://doi.org/10.1080/0144287 2.2017.1339239.

Weible, C. M., & Ingold, K. (2018). Why Advocacy Coalitions Matter and Practical Insights about Them. *Policy and Politics, 46*(2), 325–343.

Weible, C. M., Sabatier, P. A., & Pattison, A. (2010). Harnessing Expert-Based Information for Learning and the Sustainable Management of Complex Socio-ecological Systems. *Environmental Science and Policy, 13*, 522–534.

Zafonte, M., & Sabatier, P. A. (1998). Shared Beliefs and Imposed Interdependencies as Determinants of Ally Networks in Overlapping Subsystems. *Journal of Theoretical Politics, 10*(4), 473–505.

Zhu, J.-H. (1992). Issue Competition and Attention Distraction: A Zero-Sum Theory of Agenda-Setting. *Journalism and Mass Communication Quarterly, 29*(4), 825–836.

6

What Drives the Formation and Maintenance of Interest Coalitions in Water Governance Forums?

María Mancilla García and Örjan Bodin

Introduction

The literature on water governance has long acknowledged that the management and use of water resources involves a multiplicity of stakeholders with different views and interests on the resource. The worldwide set-up of water governance deliberative forums, that is, forums bringing together a diversity of stakeholders to engage in discussions and make decisions about the management of water, was driven by the hope that a deliberative space will facilitate negotiations to make different interests compatible (Weible and Sabatier 2005). One question that therefore arises is whether this expectation is met. Do actors join participatory forums to freely engage in deliberative discussions, or do they primarily engage by coordinating their positions in different matters within coalitions of interest? (Henry et al. 2011; Fischer 2014). Coalitions of interest are aggregations of a diverse set of actors (Sabatier and Jenkins-Smith 1999)

M. Mancilla García (✉) • Ö. Bodin
Stockholm Resilience Centre, Stockholm University, Stockholm, Sweden
e-mail: maria.mgarcia@su.se; orjan.bodin@su.se

© The Author(s) 2020
M. Fischer, K. Ingold (eds.), *Networks in Water Governance*, Palgrave Studies in Water Governance: Policy and Practice, https://doi.org/10.1007/978-3-030-46769-2_6

that might arise out of the need to strengthen positions—negotiation capacity can increase when a group, instead of an individual, defends the same position—or that might be the result of long-standing interest alignment. In this chapter, we use social network analysis to study individual actors' relations with others, and we conceive these relations as the basis for the constitution and maintenance of coalitions. Studying whether coalitions are constituted in participatory forums and what factors underpin these coalitions helps to provide insights on whether the forums effectively manage to promote deliberation among all actors involved, or whether deliberation rather happens among those sharing similar interests.

Through a combination of network and qualitative analysis, this chapter investigates with whom participants in a basin-level water governance forum in Brazil have established collaborative relationships, and what are the factors explaining those relationships. To this end, we focus on two methods of network analysis: Quadratic Assignment Procedure (QAP), which calculates the level of correspondence between two networks, and Exponential Random Graph Models (ERGMs), which investigates the effect of specific factors in actors' tendencies to form social links with others. Each of these methods provides different entry points to explore possible reasons why actors form links with others, as we explain in more details below. This study thus provides an analytical approach centered at the choices made by individuals in investigating if, and if so; what factors underpin the formation of coalitions? We decided to focus on coordination as the type of social relationship that underpins the potential existence of coalitions, since actors belonging to the same coalition are expected to coordinate positions within their group (Sabatier and Weible 2007). In this way, we can test whether a range of factors—such as coming from the same region or sharing the same opinion in regard to a policy—explains why actors potentially form coalitions of interests.

We develop a series of hypotheses based on the Advocacy Coalition Framework (ACF) and on Resource Dependency Theory (RDT) (Henry 2011; Calanni et al. 2015). The literature on environmental governance has made extensive use of the Advocacy Coalition Framework (hereafter ACF) to try and disentangle the role that coalitions play in determining the environmental agenda (Weible and Sabatier 2005; Hysing and Olsson

2008; Matti and Sandström 2011; Weible et al. 2019). Coalitions can be defined as groups of actors who share common interests and act based on those interests. Albeit called a "framework", the ACF has had many theoretical developments, which has led some scholars to distinguish the ACF from the ACT—Advocacy Coalition Theory (Schlager 2007; Koebele 2019b). Importantly, the ACT hypothesizes that actors form coalitions with others with whom they share what the ACF calls "deep core beliefs", which are "fundamental normative values and ontological axioms" (Sabatier 1988; Jenkins-Smith et al. 2018). The ACF also distinguishes other beliefs, called "policy core beliefs", that is, beliefs on how society and government should be organized, as also playing a role in maintaining such coalitions (Jenkins-Smith et al. 2018). Recent research has argued that core beliefs might play a less important role than previously thought (Ingold and Fischer 2014; Fischer and Sciarini 2016) and that actors tend instead to join groups for strategic reasons (Matti and Sandström 2011) and/or based on resource availability and trust vis-à-vis others (Calanni et al. 2015). Some of these findings have been theorized drawing from Resource Dependency Theory (RDT), emphasizing the need and desire of actors to get access to various kinds of resources in order to pursue their goals and objectives (Pfeffer and Salancik 1978; Casciaro and Piskorski 2005).

In this chapter, we disentangle the role of different components of beliefs to investigate what specific aspects might drive actors' decisions on coordinating with others. We also consider other factors—besides beliefs—that drive actors to coordinate with each other. By studying the social networks that actors create when they choose to coordinate with certain others, we infer the factors underpinning coalition formation and maintenance.

Our research also calls attention to the importance of context when studying coalitions and brings a Latin American case to the debate. Pierce et al. show in their recent literature review of applications of the ACF during the years 2007–2014 that while it has been applied all over the world, the majority of studied cases are situated in the Global North (2017). The Global South and particularly Latin America are underrepresented (they only identify uses of the framework in Chile). The same pattern was revealed in a previous review by Weible et al. where they showed that the United States and Europe dominated the areas of application of the framework (2009). Our study focuses on the policy subsystem of water basin governance in the south-west of Brazil, thereby contributing to a more geographically widespread use of the ACF. Further, bringing in a case from the Global South allows investigating context-sensitive elements that might help to further explain the functioning of coalitions across policy subsystems.

We studied a water basin forum that brings together more than 50 participants legally classified as belonging to one of three categories (sectors): government entities, civil society, and private users. It is thus anticipated in the design of the committee that these groups will have different views and interests, and the legislator's intention is to ensure that each of these groups is represented in the debates taking place in the forum plenary and working groups. The particular forum that we study manages a "federal river", as rivers that cross the borders of several federal states are called in Brazilian law. The forum brings together representatives of three states, namely São Paulo, Rio de Janeiro, and Minas Gerais. An equal number of representatives from each state is required by design. Additionally, three representatives from the national level participate in the forum. Our chapter seeks to answer if and to what extent ACT and RDT theories can explain why people establish coordinating links with each other in the studied forum:

1. The role of policy core beliefs, which we study by distinguishing in particular two components: policy core vision (orientation and priorities) for the basin, and policy core interests (whose welfare matters more)
2. The role of secondary beliefs, which involve the instrumental means to achieve specific policy preferences and resource allocation
3. The role of access to resources, which Resource Dependency Theory argues is a key element in explaining why people decide to link-up with others

Paying attention to the reasons why actors establish links with others will help to shed light on what explains the potential existence of coalitions of interests in water governance forums. This will also contribute to unravel the role of power differentials in constituting such coalitions and thus in affecting the participatory dynamics within the forums. In the following section, we discuss the Advocacy Coalition Framework, Advocacy Coalition Theory, and Resource Dependency Theory from which we draw for our theoretical framework and hypotheses. Then we explain the chosen methods for data collection and analysis, and the case settings. We proceed to presenting the results of the quantitative analysis which we discuss with support of our qualitative data in the last section of the chapter before the conclusion.

Theory

This chapter draws from the Advocacy Coalition Framework (ACF) and its associated theoretical developments (ACT). The chapter uses and refines part of the ACF conceptual toolset to investigate the reasons why actors enter in specific relationships (Sabatier 1998; Sabatier and Jenkins-Smith 1999; Weible 2005; Matti and Sandström 2011). The ACF sets its focus on the study of three inter-related aspects of the policy process: advocacy coalitions, policy learning, and policy change, and formulates hypotheses for each of these aspects. In this chapter, we focus specifically on the aspects that deal with the factors behind the establishment of advocacy coalitions.

The ACF distinguishes three levels of beliefs: deep core beliefs, policy core beliefs, and secondary beliefs. Deep core beliefs and policy core beliefs are seen in the ACF as rather abstract and unlikely to change—the former relates to world views, and the latter to the organization of society, government, and the economy while tied to the specific policy subsystem under study. According to the ACT, coalitions are based on the first two types of beliefs: when actors share them, they constitute stable long-term coalitions. In this study we choose to focus on the so-called "policy core beliefs" for their established importance in the literature and on the "secondary beliefs" because they have been largely neglected, judged as less important. Both types of beliefs are linked to the policy subsystem—in this case, water governance at the basin level—and therefore have territorial and topical components. Policy core beliefs are defined as those beliefs that "reflect basic orientation and value priorities for the policy subsystem and may identify whose welfare in the policy subsystem is of greatest concern" (Jenkins-Smith et al. 2018, p. 140). The ACF distinguishes this type of belief from "deep core beliefs" which are not policy specific, that is, could be applied to any policy subsystem. As Koebele (2019b, p. 44) puts it:

> *Policy core beliefs, the central element in the ACF's three-tiered hierarchical belief structure, are essentially applications of an actor's broad ontological beliefs to the bounds of a policy subsystem, making them a particularly pertinent level of belief around which actors coalesce.*

In our investigation, we will disentangle different components of "policy core beliefs" and argue that it is important to investigate the specific roles different components play, thereby challenging the ACF/ACT hypothesis that beliefs tend to be coherent. In particular, we distinguish a first component of belief that we will call policy core vision, that is, "basic orientation and value priorities for the subsystem" (Jenkins-Smith et al. 2018, p. 140). What we have called policy core vision can be seen as close in certain respects to deep core beliefs, as it can include normative values and ontological axioms (Sabatier and Jenkins-Smith 1999, p. 117). However, policy core vision is still a component of policy core beliefs because it is tied to the policy subsystem and deals with "orientation on basic value priorities", which are part of the policy core (Sabatier and Jenkins-Smith 1999).

A second component of belief is what we have summarized under policy core interests, that is, "identify whose welfare in the policy subsystem is of greatest concern", and includes "overall assessments of the seriousness of the problem, basic causes of the problem and preferred solutions for addressing the problem". "Preferred solutions for addressing the problem" are the so-called core policy preferences, a constitutive part of policy core beliefs (Sabatier and Jenkins-Smith 1999). In the ACF/ACT, these different aspects are conflated under "policy core beliefs" to which all of them contribute.

Finally, we also explore the role of secondary beliefs, which the ACF defines as dealing "with a subset of the policy subsystem or the specific instrumental means for achieving the desired outcomes outlined in the policy core beliefs" (Jenkins-Smith et al. 2018). It is also under secondary beliefs that preferences in terms of budget or resource allocation are placed.

Research not based on ACF has, among other things, focused on the role of strategic reasons behind the formation of coalitions, such as associating with particularly powerful actors over specific issues to obtain progress in one's agenda, which has been analyzed as part of Resource Dependency Theory (Pfeffer and Salancik 2003; Calanni et al. 2015). This research argues that (core) beliefs are not the most important factor in explaining how people coordinate. While beliefs might explain long-term affinities, these might not drive actors to coordinate with certain others. Recent research has also sought to associate the ACT with collaborative governance theory, arguing that "actors will coordinate for a variety of reasons other than holding shared beliefs" (Koebele 2019b, p. 45). Our chapter contributes to this literature by nuancing the ACT hypotheses and engages in the debate on whether RDT is a more appropriate or complementary explanation as to why people form coordinating links with others.

We formulated a series of hypotheses to test patterns of collaborative links in the network of actors participating in a water basin forum in Brazil. If coalitions exist, they would be influential in affecting deliberative behaviors only if their constituting members coordinate their activities. Thus, we assume that if coalitions exist, they would entail patterns of coordination that coincide with the coalitions themselves.

Hypotheses

Our hypotheses are complementary, and not contradictory, nor mutually exclusive. Finding support for one or all of them allows us to assess the factors that might drive participants in the forum to establish coordinating links with others. Our qualitative data helps us to nuance and further investigate our quantitative findings.

ACF hypothesizes that policy core beliefs play a key role in driving actors to link to each other and constitute coalitions. We are specifically interested in assessing whether the component of policy core belief that we have called "policy core vision" leads indeed to the constitution of coalitions. Therefore, if any two given actors share a vision of the basin, they would coordinate with each other. So, drawing on the ACF, we formulated our base-line hypothesis:

H1 *Actors primarily coordinate with those who share their vision of the basin*

Secondly, we tested whether we could identify links between actors based on them belonging to the same sector (i.e. government entities, civil society, and private users), which we used as a proxy to assess whether actors shared policy core interests, the second component of policy core beliefs that we were interested in disentangling:

H2 *Actors tend to coordinate with others from their sector*

The constitution of sectorial groups in the policy subsystem is created by law, suggesting that belonging to the same sector typically leads to converging interests based on how government should be structured and whose interests should prevail. The ACF specifically identifies "proper distribution of authority between government and markets" and "proper distribution of authority among levels of government" as empirical precepts of the policy core beliefs (Sabatier and Jenkins-Smith 1999), and it is these aspects that we aim at untangling with the proxy of the sector.

Through our third hypothesis, we assessed whether secondary beliefs could help us explain links, by using the state attribute as a proxy. This hypothesis goes against the ACT theoretical core which tends to consider secondary aspects as less important than policy core aspects. Indeed, among the ACT hypotheses we find:

> *Actors within an advocacy coalition will show substantial consensus on issues pertaining to the policy core, although less so on the secondary aspects.*

and

An actor (or coalition) will give up secondary aspects before acknowledging weakness of the policy core. (Jenkins-Smith et al. 2014)

We were interested in investigating whether secondary beliefs were important in explaining why actors establish links with others and therefore explain patterns of coordination. If this were the case, it would suggest that actors, contrary to what the quote affirms, don't find secondary aspects negligible and easily abandoned, but instead find them important. If secondary aspects, instead of policy core aspects, influence the way people establish coordination links, it would mean that such aspects are key in determining behavior and coalition formation. To test whether this holds true in our case, we formulated the following hypothesis that contradicts the assumption that secondary beliefs are not important in forming collaborative relationships:

H3 *Actors tend to coordinate with others from their state*

The classification into different states is not based on an actor's choice, but rather on a geographic reality. We found that the state was a good proxy to assess some of the aspects the ACF considers part of secondary beliefs such as "seriousness of specific aspects of the problem in specific locales", "importance of various causal linkages in different locales and over time", "decisions concerning administrative rules, budgetary allocations, disposition of cases" and so on, and "information regarding performance of specific programs or institutions". All these aspects are strongly linked to different geographic realities and we thus considered the state proxy adequately captures them.

Then, we evaluated whether people tended to coordinate with those who were perceived as most influential. Influence perception has been used as a way to assess actors' perception of others' access to resources (Fischer and Sciarini 2015; Ingold and Leifeld 2014). Within Resource Dependency Theory in particular, this measure has been used to assess the effect of access to diverse resources, including economic or social resources. Indeed, influence-perception reflects actors' assessment on others' capacity to effectively translate their beliefs into actions, be it because they have the social status to do so or the capacity to fund financially those actions (Henry 2011). Focusing on influence perception instead of on measures of resource-access such as formal authority or financial resources allows one to assess the reasons actors have to coordinate with others in a broader sense. Actors might have access to a set of resources that are difficult to assess or even not measurable—such as social status or prestige. Moreover, actors might not be able to assess the financial capacity of others in such large forums as the one we focus on here. Finally, in cases in the Global South where institutional capacity is often weak, formal authority might not be perceived as necessary leading to action (Abers and Keck 2009). Therefore, by investigating whether actors tend to coordinate more with those perceived as more influential, we investigate the relevance of the Resource Dependency Theory for this particular case.

H4 *Actors tend to coordinate more with those perceived as more influential*

This hypothesis tries to assess the claim in the literature that actors tend to constitute coalitions for strategic reasons, as identified by Resource Dependency Theory, rather than based on core policy beliefs.

Methods

Data Collection

In water governance settings, defining the study object can be challenging since water does not adhere to administrative borders or other human-defined boundaries. In our case, we chose to focus on the forum with legal competencies over the management of a river, the Paraíba do Sul river in Brazil. The actors included in our quantitative analysis consist of the participants in such a forum, that is, the plenary of a water basin committee. The study population was further refined based on one additional condition: an actor having attended at least two of the last six plenary meetings. This resulted in a population of 45 people (actors), of which 3 declined to participate in the study, which gives us a response rate of 93%.

We presented respondents with the list of all participants in the plenary of the committee and asked them to evaluate each participant following three criteria: (1) whether they saw them as influential, whether (2) they coordinated positions with them—which we defined as "some degree of working together to achieve similar policy objectives" inspired by Sabatier and Weible (2007, p. 196), and gave as examples coordination in voting behavior and support in deliberations—and (3) whether they believed they shared the same vision of the basin. The first question evaluated influence on a five-point scale where 1 was "not at all" and 5 "completely". The last two questions were rated on a four-point scale where: 1 means "rarely or never", 2 means "sometimes", 3 means "often" and 4 means "constantly". For example, if participants coordinated on how to vote on every decision with another participant, they would give a 4 in terms of coordination with that participant. For both scales, 0 was used to indicate that the respondent did not know that participant. Perceived influence of any given participant was assessed based on the average rating from all other participants.

To complement this data, we ran extensive semi-structured interviews with participants in the committee's plenary during which they could speak about their general views on the system of management, its problems, the progress made in the last few years, as well as the vision ahead. We additionally interviewed some public officers at the executive agency of the committee, as well as former participants in the plenary and actors who had been involved in the set-up of the committee. The total amount of time spent with each interviewee individually ranged from one to three hours. Finally, plenary committee meetings, those of the technical chamber, some of those of other forums in which participants in the committee's plenary also participated were attended, during which notes were taken.

Data Analysis

All qualitative data was imported into the software NVivo11 and coded following an abductive thematic analysis (Boyatzis 1998). A summary of each interview was first drafted and in a second step, quotes relating to "coalitions" were identified across data sources. The data coded under "coalitions" was then re-examined by the researchers to find explanations as to why coalitions were constituted, instances of communication or collaboration between coalitions, and instances of breaches to any given coalition's collective interest by one of its members. The objective of the analysis of qualitative data was to achieve a nuanced perspective of different possibilities at play concerning coalition behavior, that is, if there were specific reasons mentioned—and what those were—to coordinate positions within specific groups.

Quadratic Assignment Procedure

We constructed two networks based on the responses to the question on the vision of the basin and on coordination. To assess whether core policy belief homophily—that is, sharing beliefs with an actor—affected coordination patterns we used a Quadratic Assignment Procedure (QAP) in UciNet (see Borgatti et al. 2002). QAP is a social network analysis tool, frequently used to investigate policy networks, that calculates the level of correlation or association between two network matrices (Dekker et al. 2007; Lubell et al. 2012). QAP, as a nonparametric technique relying on simulations to estimate levels of statistical significance, takes into account that network data is rarely independent by preserving the observed distribution of links in the network while estimating probabilities. In our case, it allowed us to test whether the vision of the basin of a particular actor aligns with the vision of those with whom the actor coordinates. It measures how strong the correlation is, and how likely it is that such correlation is not random.

We considered that if actors formed coalitions based on their policy core-beliefs, we would observe a perfect correlation of the coordination network and the shared-vision network. These data explicitly sought to uncover how people perceived others' understandings of the nature of the basin as well as their sharing of the normative vision of what the basin ought to be, that is, beliefs about what the basin is and should be, for example, a commodity, a cultural good, and so on.

Hypothesis 1 is our baseline hypothesis. To further evaluate other possible factors behind link formation, which we cover through hypotheses 2–4, we turned to Exponential Random Graph Models (ERGM) (Lusher et al. 2013).

Exponential Random Graph Models

Exponential Random Graph Models (ERGM) are commonly used in network analysis to investigate what factors could explain the tendency actors have to link to certain others. This is accomplished by investigating the prevalence of certain building blocks (configurations) in a network (see Box). ERGMs allow testing the prevalence of several of those configurations simultaneously. These configurations can be purely structural, such as when incoming links are reciprocated, and/or they can be based on certain node attributes (e.g. actors of a certain type tend to have more links than others).

Technical Specificities of ERGM:

ERGM uses maximum likelihood simulation techniques to fit a parameter vector θ to a stochastic network model (Lusher et al. 2013):

$$P^{\theta}\left(X = x\right) \propto \exp\left\{\theta s\left(x\right)\right\}$$

where X is a random network (x is the empirical network), and $s(x)$ is a known vector of building blocks (configurations) on x.

A well-fitting ERGM would then adequately represent the network through a set of configurations and their associated coefficients. The coefficients capture if a certain configuration is prevalent, suppressed, or neither (the latter means the coefficient would not be significantly different from zero). Yet, fitting an ERGM can suffer from convergence problems. A Goodness of Fit test is thus normally carried out to find out if the model is suffering from convergence problems.

The main difference between ERGM and "off-the-shelf" statistical methods such as linear regression is that ERGM does not assume independence of data observations (Cranmer and Desmarais 2011). A network is created through patterns of connections, and thus, implies data interdependency.

To test our hypotheses 2–4, three types of node (actor) attributes were explicated using ERGM: whether people interacted more when they belonged to the same state (which we tested for each of the states); whether they interacted more if they belonged to the same sector (which we tested for each of the sectors); and whether people tended to link more with those they perceived as more influential.

We also included two control configurations, that is, configurations commonly occurring in social networks that could influence the results of the ERGM estimations (if not properly taken into account). The first being if people from certain states and sectors tended to receive, and/or send, more ties than others from other states or sectors. The second was reciprocity, that is, we controlled for the commonly occurring tendency for people to reciprocate incoming social ties (Robins 2015).

Case

The policy system on which we focused is the management system of the basin of the Paraíba do Sul river, which flows through the states of Rio de Janeiro, Minas Gerais, and São Paulo (see Fig. 6.1). The Paraiba do Sul river covers a basin area of 56,500 km^2 and provides water for 17.5 million people. The main water uses are water provision, sewerage (dilution of used waters), irrigation, and hydroelectricity generation. Also, although less important, the river is used for fishing, aquaculture, and tourism. The river counts with a federal committee that deals with the whole extension of the basin, on which we focus here and with seven smaller committees that cover the different portions and estuaries of the Paraíba do Sul River.

The federal basin committee of the Paraíba do Sul river has overarching responsibility over the management of the system. Among its main competencies, we find the definition of the quality of the river's water, of the rights of use, of the values for payment of water use, as well as the

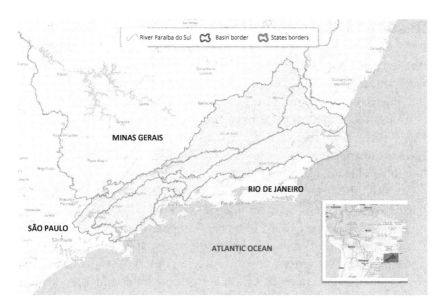

Fig. 6.1   Water Basin of River Paraíba do Sul (Mancilla García and Bodin 2019)

approval and implementation of the Water Resources Plan for the basin. The CEIVAP, which is an acronym for "Committee for the Integration of the Paraíba do Sul River", was created in 1996. Its statutes define that 40% of plenary members represent the users (industries, hydroelectric companies, agriculture, provision companies, etc.); 35% represent governmental entities at the federal, state, and municipal levels; and 25% represent civil organizations (associations, NGOs, universities). The representatives of these three categories are equally distributed between the three states. Besides the plenary, the CEIVAP also counts with a technical chamber—composed of six members per state, two of each sectorial category—and with several working groups. Additionally, the plenary elects a three-member directorate for two years with one representative from each category, and each of them from a different state. Representatives of government entities hold the presidency, which rotates between the three states.

Results of the Quantitative Analysis

The results of our QAP on the correlation between the coordination network and the vision of the basin network gave a Pearson's coefficient of 0.717 (p<0.000). This demonstrates that although the vision of the basin is strongly correlated to patterns of coordination, there is still variability between the two. Therefore, the reasons explaining coordination merit further investigation, as sharing the vision of the basin is not the single unequivocal factor.

We used the freely accessible software MPNet (Peng et al. 2009) for our ERGM.[1] The results are summarized in Table 6.1.

When interpreting ERGM results, it is important to keep in mind that it is the combination of what is significant and what isn't that allows us to find support for certain hypotheses and reject others. As explained above, we focus on interactions based on sector and state, and thus include sender and receiver configurations for these attributes as controls. The results indicate that the private sector is less active than the other two sectors (the parameter estimate for receiving links is significantly negative), but when they link, they link more with others from their own sector (the interaction configuration in Table 6.1 is positive and significant). Thus, private actors appear as more coherent than the other actor groups at the network level. Note that belonging to the same sector is not a significant predictor of links when the sector is the public sector or civil society. It is only significant for the private sector. By contrast, belonging to the same state is a significant predictor of links for all three states.

To evaluate the validity of the model, Goodness of Fit (GOF) tests need to be run. This allows us to assess whether the model is converging properly (see Box). The GOF test also allows us to see if the ERGM is able to adequately reproduce configurations not explicitly included in the model. The results of the GOF (see Appendix) confirmed that the model was not able to fully capture all possible configurations, most notable triadic closure (i.e. three nodes all being interconnected in various ways). We acknowledge that our model, which is derived directly from our hypotheses, is a bit simplistic and hence doesn't capture these more

---

[1] The software and manual can be found at: http://www.melnet.org.au/pnet/.

Table 6.1 ERGM results

Configuration[a]	Parameter	Standard Error
Predictions of interest		
State MG_InteractionA	1.5869[c]	0.281
State RJ_InteractionA	0.6684[c]	0.286
State SP_InteractionA	1.8125[c]	0.284
Public Sector_InteractionA	0.0313	0.318
Private Sector_InteractionA	0.6110[c]	0.260
Civil Society_InteractionA	0.4187	0.326
Influence_SenderA	0.0004[c]	0.000
Influence Receiver_A	0.0007[c]	0.000
Controls		
State SP_SenderA	−0.3412	0.221
State SP_ReceiverA	0.0355	0.230
State RJ_SenderA	−0.0615	0.225
State RJ_ReceiverA	0.4647	0.233
Public Sector_SenderA	−0.4087	0.228
Public Sector_ReceiverA	−0.0703	0.216
Private Sector_SenderA	−0.3193	0.224
Private Sector_ReceiverA	−0.6023[c]	0.234
Link[b]	−3.1468[c]	0.370
Reciprocity	0.7468[c]	0.211

[a]the send and receive configurations used the state MG and the sector civil society as baselines, that is, all parameters are estimated in relation to these categories
[b]this configuration represents the likelihood for the presence of a link, thus it captures the network density. For example, a parameter value of 0 would mean that there is a 50% probability for a link to exist between any pair of nodes
[c]indicates significance

complex configurations so well. Even though we acknowledge this could potentially affect some of our results, we deem it being unlikely there would be any major overall differences. This was supported by conducting a rather extensive exploration where more complicated configurations were included. This exploration revealed it (1) was very difficult to include additional configurations due to convergence problems (see Box), and (2) in those few cases, the model actually converged, most of our results remained the same (the interaction configurations for all states were always significantly greater than 0, albeit in regards to sector, there were instances where the civil society interaction configuration became significantly greater than 0).

Discussion

The results from the QAP support our first hypothesis, that is, that the "vision" aspect of policy core beliefs is important in explaining why actors linked to each other. This result supports the traditional view of ACT, in which policy core beliefs play a central role in leading to and explaining the maintenance of coalitions (Jenkins-Smith et al. 2014).

The result of the QAP does not, however, indicate a perfect correlation between sharing the vision of the basin and peer-to-peer coordination. This suggests that other factors also contribute in driving coordination between participants in the basin. Moreover, during the interviews, some participants declared that they felt unable to evaluate the vision of all others, since they associated the vision with ontological axioms and normative views and they were not sure they knew all their colleagues well enough to be able to assess that. Yet, some of them indicated that those very same people, for whom they could not evaluate the vision of the basin, could be their allies in terms of setting the priorities of actions to undertake. Therefore, our qualitative data suggests that policy core vision is not the only reason explaining links. Instead, there seem to be several reasons why actors associate with others, supporting recent research that also points to that direction (Koebele 2019a, b).

The ERGM results indicate that H2 is only partially supported. Indeed, only private sector participants tend to link more with each other than with participants belonging to other sectors. This was not the case for either public sector or civil society participants. This is interesting because it suggests that certain aspects of policy core beliefs, namely, what we have called here policy core interests, don't play an important role in explaining coordination. This means that for many actors, sharing specific aspects of their policy core beliefs, namely those associated with belonging to the same sectors, did not significantly influence their choices on with whom to coordinate. By contrast, for certain others, in our case members of the private sector, it seemingly had an influence; thereby confirming the ACT hypothesis that policy core interests (in addition to policy core vision) are also important in explaining the constitution of coalitions, albeit only for this specific sector.

As mentioned above, we observe that belonging to the same state is a good predictor for linking with others. This supports H3, and suggests that belonging to the same state might be a reason underpinning the constitution of coalitions. Our qualitative data also supports this finding. As we were interviewing all participants in the committee, some of them recommended that we find the most influential people in each state and interview them only, since, we were told, that would be enough to understand how the system worked. One of our interviewees who had been involved in diverse aspects of water management in the region for several decades provided an interesting answer to our question "In your view, what are the characteristics of the basin?" For this actor, one could not speak of characteristics of the basin as a whole, but rather of characteristics of the basin in each of the states, thereby indicating that some actors perceive water governance as being a state-level affair and not a basin-level affair.

All through our qualitative data, we find instances of the states being perceived as a collective actor whose interests should be defended by those belonging to each of them. Another example of this is visible in the narrations of the drought crisis of the years 2014–2015, when it was decided that some water of the Paraíba do Sul would be deviated to one of the reservoir systems in São Paulo. This was controversial because the river has historically been used as a source of drinking water for the state of Rio de Janeiro, and it was feared that the works to deviate part of the water to São Paulo would affect the provision in Rio. These elements point to the existence of a commonly shared vision of state interest as a common interest in which it is important that the state one belongs to gets the most resources—funds, water, and so on—or at least does not lose in exchange with other states. Previous research has already identified the importance of geography in the constitution of coalitions (Koebele 2019a).

The joint analysis of H2 and H3 shows that participants tend to coordinate with others within their state, except for private sector representatives that also unite regardless of state. Since the private sector confirms the ACT theoretical predictions, we choose to focus on elaborating why civil society members and the public sector seemingly deviate from what ACF/ACT predicts (i.e. that policy core *interests* are more important than secondary beliefs). This result thus contributes to the literature nuancing the importance of core beliefs and suggesting that different reasons might explain why actors link to each other (Matti and Sandström 2011; Ingold and Fischer 2014; Calanni et al. 2015; Fischer and Sciarini 2016).

The public sector brings together representatives from municipalities and from state-level institutions. There are also participants who represent the national state, but these are very few (three formally, of which two were included in our sample). Hence, the group composed of public actors is actually a quite diverse group. Our results suggest that this diversity makes them less inclined to coordinate with each other simply because they are public actors (thereby sharing policy core interests).

We further investigate the case of civil society with the help of our qualitative data. Considering the nature of the interests defended in civil society, one could expect that civil society representatives would pay more attention to whether an actor was part of this sector or not when choosing with whom to coordinate. Civil society members bring together representatives from NGOs, local associations, and universities, which tend to share a common interest for environmental issues (and thus can be said to somewhat take on the role as the safeguards of the natural environment). However, this common interest does not necessarily apply to all of them. Some of the associations represented, and some university representatives, might be more inclined toward defending water use efficiency rather than the environment. Thus, akin to the public sector, the actors in this sector were also rather heterogeneous.

Nonetheless, it remains somewhat surprising that civil society does not represent a more cohesive coalition across states. Yet, our qualitative data provides us with examples of tension between different sectors within the same state, which suggest that state interest is sometimes contested. One such example is the debate around the water payment by committee Guandu, which is one of the committees within the state of Rio de Janeiro. While there were transposition works affecting the Paraíba do Sul since the beginning of the twentieth century, 1952 marked an important date in the history of the river management because it is when the works of the Santa Cecilia plant were finished. This allowed a transposition of waters from the Paraiba do Sul to the Guandu river up to $160 \text{ m}^3/\text{s}$. Originally, the transposition was done to support the generation of electricity. Yet, the new outflow of river Guandu led it to become a key source of provision of water for human use for the metropolitan region of Rio de Janeiro (Teixeira 2010). The committee Guandu, a state-based committee in Rio de Janeiro, pays a certain amount to the federal committee CEIVAP to compensate for the transposition. During our time in the field, there was a debate about raising the amount that the committee Guandu should pay. Some of the participants in CEIVAP felt that the payment was insufficient, and argued that a higher payment would help support environmental restoration activities in the basin. During the debates, some of the civil society representatives supporting this position were from the state of Rio de Janeiro. Other members of this state expressed their surprise in regard to the attitude of civil society members that, in their view, were more worried about the environmental state of the river outflow—across the three states—than about the drinking water of their fellow state inhabitants. Hence, in the view of certain other actors, civil society members paid insufficient attention to which state they belonged. Civil society members instead considered that guaranteeing a sufficient outflow for the river would allow for long-term uses, such as water provision. This example shows that at times, the state interest is contested.

Commenting more generally on existing tensions between different groups within the plenary, one of our interviewees explained that the outcome of such tensions could have diverse resolutions. It could be that the groups vote together based on what they believe in, but this respondent also told us that in certain cases, exchanges of favors can determine the voting. For example, the interviewee explained, it could be that in a case of tension between what government entities considered to be in the interests of a state and what civil society would think is of concern for the environment, the government entities could commit to develop projects in certain areas of the region, to somehow satisfy some of the environmental concerns. This illustrates that, in certain cases, actors can be convinced to act in favor of the interests of the state, as understood by government entities, although their rationale for doing so is nonetheless based on their interest in environmental issues.

Finally, those civil society representatives defending the environmental interest told us that it was crucial for them to be organized and united so that they could make their voices heard, supporting the idea that setting up coalitions would give strength to their position. Moreover, those of our interviewees who were not participating in the plenary—either because they were former members of the plenary, or had been involved in the set-up of the committee—coincided in arguing that the environmental interest would be more successfully defended if civil society members managed to be better organized and more cohesive. This suggests that civil society actors are less coherent in coordinating their activities in comparison to how coherent their core policy interests are.

These examples and the joint analysis of the results of hypotheses 2 and 3 indicate that there is a variety of affinities for actors. This suggests that coalitions are at least partly overlapping and that while there might be a core base or attractor holding people together, the reasons that actually determine final behavior are dependent on the specific issue being discussed or rather the specific negotiation setting at play. This indicates that the context and the multiple contextual nuances play a role in determining concrete behavior. Therefore, combining network analysis with qualitative data helps to provide a nuanced view of how networks operate and when different links between specific actors become important in explaining behaviors.

Finally, the results of our ERGM provide firm support for hypothesis 4, that is, that influence is important in explaining why actors link up with others (both for incoming and outgoing links). This also seems coherent with the above-mentioned examples, in which we discussed how diverse types of processes are at play in determining how actors link up with others in specific cases. Therefore, we distinguish in our case multiple factors actors consider when choosing to coordinate with others. Our findings are thus different from those that focus on more stable coalitions structures and identify links across coalitions leading to win-win arrangements (Weible and Sabatier 2009). Instead, we identify instances in which actors might engage in political bargaining to pursue their own specific objectives and thus seemingly act in contradiction to the interest of their coalitions.

Conclusion

Our results suggest that participants in the committee are forming coalitions of interest, but those are not necessarily solely built on shared core policy beliefs, as the ACT has hypothesized. We used Quadratic Assignment Procedure to test whether the ACT hypothesis on core beliefs was verified in our case. This method was particularly useful as it showed that the correspondence between coalition—which we investigated through effective links between individual actors—and beliefs was strong but not absolute. The use of Exponential Random Graph Models in combination with an analysis of rich qualitative data allowed us to investigate a set of different reasons why actors link to others. The results of this investigation led us to nuance and enrich the ACT literature by disentangling the role of different elements—policy core vision, policy core interests, and secondary beliefs—and putting them in dialogue with insights from other perspectives, such as Resource Dependency Theory.

Our analysis shows that there might be different reasons for establishing links with others, constituting diverse coalitions—such as state-based or interest-based as we saw it was the case of the private sector. Moreover, actors might not necessarily always be loyal to their coalitions. This implies that actors do not only belong to one specific coalition; rather coalitions tend to overlap in scope and interests, and reasons for coordination can be adversarial but don't have to be. In such setting, secondary beliefs do seem to play a more crucial role than do core policy interests, which contradicts what the ACF/ACT postulates. More research is needed to determine when and why people choose diverse aspects of their belief systems to align with others and under which conditions strategic reasons are prevalent. Emphasizing the specifics of the local context when applying broad theoretical perspectives such as ACF will help us to better understand the multiple reasons why actors choose to coordinate their activities with certain others, and why they choose not to.

Acknowledgements The authors would like to thank Annica Sandström for her useful comments and help in improving the chapter. We extend our gratitude to Andrea Nascetti for his help in producing the included map.

Appendix

Please refer to mpnet manual (http://www.melnet.org.au/pnet/) *for further specification on the configurations included.*

Table 6.2   Results Goodness of Fit test

Statistics[a]	Observed	Mean	StdDev	t-ratio[b]
ArcA	443.0000	442.0410	16.321	0.059
ReciprocityA	109.0000	108.6700	8.388	0.039
In2StarA	2875.0000	2886.8970	211.054	−0.056
Out2StarA	3215.0000	2494.9590	187.543	3.839*
In3StarA	14464.0000	14920.6940	1836.112	−0.249
Out3StarA	19978.0000	10058.6150	1274.225	7.785*
TwoPathA	5134.0000	5227.1030	377.396	−0.247
Transitive-TriadA	2437.0000	1910.1700	205.045	2.569*
Cyclic-TriadA	601.0000	606.3880	68.749	−0.078
T1A	76.0000	65.8910	16.210	0.624
T2A	758.0000	642.9230	121.836	0.945
T3A	1188.0000	1069.4180	154.704	0.767
T4A	666.0000	564.8670	79.575	1.271
T5A	728.0000	530.8440	76.898	2.564*
T6A	754.0000	768.1170	117.550	−0.120
T7A	2710.0000	2954.5350	307.301	−0.796
T8A	3057.0000	2662.5200	292.668	1.348
SinkA	1.0000	0.0250	0.162	6.004*
SourceA	0.0000	0.1560	0.387	−0.403
IsolateA	0.0000	0.0050	0.071	−0.071
AinSA	724.0845	720.8595	32.150	0.100
AoutSA	729.9856	718.1907	32.372	0.364
AinSA2	724.0845	720.8595	32.150	0.100
AoutSA2	729.9856	718.1907	32.372	0.364
AinAoutSA	151.0861	161.4284	1.906	−5.427*
ATA-T	778.9042	745.0362	42.059	0.805
ATA-C	699.2587	730.0923	44.200	−0.698
ATA-D	801.7768	720.1359	43.161	1.892
ATA-U	744.1016	769.7574	42.431	−0.605
ATA-TD	1580.6810	1465.1722	84.635	1.365
ATA-TU	1523.0058	1514.7937	83.959	0.098
ATA-DU	1545.8784	1489.8934	83.938	0.667
ATA-TDU	2324.7826	2234.9296	125.799	0.714
A2PA-T	2214.8368	2488.6216	86.287	−3.173*
A2PA-D	1328.5037	1167.8056	53.115	3.025*
A2PA-U	1139.4226	1361.0730	48.061	−4.612*

(*continued*)

Table 6.2 (continued)

Statistics[a]	Observed	Mean	StdDev	t-ratio[b]
A2PA-TD	3543.3406	3656.4272	134.584	−0.840
A2PA-TU	3354.2595	3849.6946	123.422	−4.014*
A2PA-DU	2467.9264	2528.8786	81.214	−0.751
A2PA-TDU	4682.7632	5017.5002	165.424	−2.024*
MG_InteractionA	68.0000	67.9590	5.888	0.007
SP_SenderA	169.0000	167.9360	9.847	0.108
SP_ReceiverA	174.0000	172.6500	10.878	0.124
SP_InteractionA	103.0000	102.3460	7.310	0.089
RJ_SenderA	126.0000	126.6080	8.435	−0.072
RJ_ReceiverA	140.0000	140.3990	8.973	−0.044
RJ_InteractionA	59.0000	59.1810	5.723	−0.032
Pub_SenderA	120.0000	119.4470	8.604	0.064
Pub_ReceiverA	143.0000	143.2610	9.217	−0.028
Pub_InteractionA	39.0000	38.7460	5.062	0.050
Pri_SenderA	196.0000	196.2160	10.594	−0.020
Pri_ReceiverA	165.0000	164.1760	10.323	0.080
Pri_InteractionA	81.0000	80.7810	7.615	0.029
CS_InteractionA	42.0000	41.9310	4.537	0.015
Influence_SenderA	778272.0000	777036.6190	31232.690	0.040
Influence_ReceiverA	868151.0000	867535.4150	31207.568	0.020
stddev_indegreeA	6.0896	6.1498	0.398	−0.151
skew_indegreeA	0.7467	0.8925	0.202	−0.723
stddev_outdegreeA	7.3259	4.3280	0.358	8.384*
skew_outdegreeA	1.2529	0.5950	0.321	2.050*
clusteringA_tm	0.4747	0.3647	0.016	6.804*
clusteringA_cm	0.3512	0.3471	0.017	0.235
clusteringA_ti	0.4238	0.3302	0.016	5.773*
clusteringA_to	0.3790	0.3822	0.019	−0.169

[a]Mahalanobis distance is 175943. This measure is similar to the standard R^2 in regression analysis, but should be as small as possible, and it is not confined to a 0–1 interval

[b]'*' means that the t-ratio is higher than 2.0, indicative of statistical significance

References

Abers, R., & Keck, M. (2009). Mobilizing the State: The Erratic Partner in Brazil's Participatory Water Policy. *Politics & Society, 37,* 289–314.

Borgatti, S. P., Everett, M. G., & Freeman, L. C. (2002). *Ucinet for Windows: Software for Social Network Analysis.* Harvard, MA: Analytic Technologies.

Boyatzis, R. (1998). *Transforming Qualitative Data: Thematic Analysis and Code Development.* Thousand Oaks, CA: Sage.

Calanni, J. C., Siddiki, S. N., Weible, C. M., & Leach, W. D. (2015). Explaining Coordination in Collaborative Partnerships and Clarifying the Scope of the Belief Homophily Hypothesis. *Journal of Public Administration Research and Theory, 25*(3), 901–927.

Casciaro, T., & Piskorski, M. J. (2005). Power Imbalance, Mutual Dependence, and Constraint Absorption: A Closer Look at Resource Dependence Theory. *Administrative Science Quarterly, 50*(2), 167–199.

Cranmer, S. J., & Desmarais, B. A. (2011). Inferential Network Analysis with Exponential Random Graph Models. *Political Analysis, 19*(1), 66–86.

Dekker, D., Krackhardt, D., & Snijders, T. A. B. (2007). Sensitivity of MRQAP Tests to Collinearity and Autocorrelation Conditions. *Psychometrika, 72*(4), 563–581.

Fischer, M. (2014). Coalition Structures and Policy Change in a Consensus Democracy. *Policy Studies Journal, 42*(3), 344–366.

Fischer, M., & Sciarini, P. (2015). Unpacking Reputational Power: Intended and Unintended Determinants of the Assessment of Actors' Power. *Social Networks, 42*, 60–71. https://doi.org/10.1016/j.socnet.2015.02.008

Fischer, M., & Sciarini, P. (2016). Drivers of Collaboration in Political Decision Making: A Cross-sector Perspective. *The Journal of Politics, 78*(1), 63–74.

Henry, A. D. (2011). Ideology, Power, and the Structure of Policy Networks. *Policy Studies Journal, 39*(3), 361–383.

Henry, A. D., Lubell, M., & McCoy, M. (2011). Belief Systems and Social Capital as Drivers of Policy Network Structure: The Case of California Regional Planning. *Journal of Public Administration Research and Theory, 21*(3), 419–444.

Hysing, E., & Olsson, J. (2008). Contextualising the Advocacy Coalition Framework: Theorising Change in Swedish Forest Policy. *Environmental Politics, 17*(5), 730–748.

Ingold, K., & Fischer, M. (2014). Drivers of Collaboration to Mitigate Climate Change: An Illustration of Swiss Climate Policy over 15 Years. *Global Environmental Change, 24*(1), 88–98.

Ingold, K., & Leifeld, P. (2014). Structural and Institutional Determinants of Influence Reputation: A Comparison of Collaborative and Adversarial Policy Networks in Decision Making and Implementation. *Journal of Public Administration Research and Theory, 26* (1), muu043. https://doi.org/10.1093/jopart/muu043

Jenkins-Smith, H. C., Nohrstedt, D., Weible, C. M., & Ingold, K. (2018). The Advocacy Coalition Framework: An Overview of the Research Program. In

C. M. Weible & P. A. Sabatier (Eds.), *Theories of the Policy Process* (4th ed., pp. 135–172). New York, NY: Routledge.

Jenkins-Smith, H. C., Nohrstedt, D., Weible, C. M., & Sabatier, P. A. (2014). The Advocacy Coalition Framework: Foundations, Evolution and Ongoing Research. In P. A. Sabatier & C. M. Weible (Eds.), *Theories of the Policy Process* (pp. 184–217). Boulder, CO: Westview Press.

Koebele, E. (2019a). Cross-coalition Coordination in Collaborative Environmental Governance Processes. *Policy Studies Journal.* https://doi.org/10.1111/psj.12306.

Koebele, E. (2019b). Integrating Collaborative Governance Theory with the Advocacy Coalition Framework. *Journal of Public Policy, 39*(1), 35–64.

Lubell, M., Scholz, J., Berardo, R., & Robins, G. (2012). Testing Policy Theory with Statistical Models of Networks. *Policy Studies Journal, 40*(3), 351–374.

Lusher, D., Koskinen, J. H., & Robins, G. (2013). *Exponential Random Graph Models for Social Networks: Theory, Methods, and Applications.* Cambridge, UK: Cambridge University Press.

Mancilla García, M., & Bodin, Ö. (2019). Participation in Multiple Decision Making Water Governance Forums in Brazil Enhances Actors' Perceived Level of Influence. *Policy Studies Journal, 47*(1), 27–51. https://doi.org/10.1111/psj.12297

Matti, S., & Sandström, A. (2011). The Rationale Determining Advocacy Coalitions: Examining Coordination Networks and Corresponding Beliefs. *Policy Studies Journal, 39*(3), 385–410.

Peng, W., Robins, G., & Pattison, P. (2009). *PNet: Program for the Simulation and Estimation of Exponential Random Graph Models.* Melbourne School of Psychological Sciences, The University of Melbourne.

Pfeffer, J., & Salancik, G. R. (1978). *The External Control of Organizations: A Resource Dependence Perspective.* New York, NY: Harper and Row.

Pfeffer, J., & Salancik, G. R. (2003). *The External Control of Organizations: A Resource Dependence Perspective.* Stanford, CA: Stanford Business Books.

Pierce, J. J., Peterson, H. L., Jones, M. D., Garrard, S. P., & Vu, T. (2017). There and Back Again: A Tale of the Advocacy Coalition Framework. *Policy Studies Journal, 45*, 13–46.

Robins, G. (2015). *Doing Social Network Research: Network-Based Research Design for Social Scientists.* London, UK: Sage.

Sabatier, P. A. (1988). An Advocacy Coalition Framework of Policy Change and the Role of Policy-oriented Learning Therein. *Policy Sciences, 21*, 129–168.

Sabatier, P. A. (1998). The Advocacy Coalition Framework: Revisions and Relevance for Europe. *Journal of European Public Policy,* 5(1), 98–130.

Sabatier, P., & Jenkins-Smith, H. (1999). The Advocacy Coalition Framework: An Assessment. In P. A. Sabatier (Ed.), *Theories of the Policy Process* (pp. 117–166). Boulder, CO: Westview Press.

Sabatier, P. A., & Weible, C. M. (2007). The Advocacy Coalition Framework: Innovation and Clarification. In P. A. Sabatier (Ed.), *Theories of the Policy Process* (pp. 189–220). Boulder, CO: Westview Press.

Schlager, E. (2007). A Comparison of Frameworks, Theories, and Models of Policy Processes. In P. A. Sabatier (Ed.), *Theories of the Policy Process* (2nd ed., pp. 293–321). Boulder, CO: Westview Press.

Teixeira, V. L. (2010). Universidade do Estado do Rio de Janeiro Centro de Tecnologia e Ciência Instituto de Química. Universidade do Estado do Rio de Janeiro.

Weible, C. M. (2005). Beliefs and Perceived Influence in a Natural Resource Conflict: An Advocacy Coalition Approach to Policy Networks. *Political Research Quarterly,* 58(3), 461–475.

Weible, C. M., Ingold, K., Nohrstedt, D., Henry, A. D., & Jenkins-Smith, H. C. (2019). Sharpening Advocacy Coalitions. *Policy Studies Journal.* https://doi.org/10.1111/psj.12360.

Weible, C. M., & Sabatier, P. A. (2005). Comparing Policy Networks: Marine Protected Areas in California. *Policy Studies Journal,* 33(2), 181–201.

Weible, C. M., & Sabatier, P. A. (2009). Coalitions, Science, and Belief Change: Comparing Adversarial and Collaborative Policy Subsystems. *Policy Studies Journal,* 37(2), 195–212.

Weible, C. M., Sabatier, P. A., & McQueen, K. (2009). Themes and Variations: Taking Stock of the Advocacy Coalition Framework. *Policy Studies Journal,* 37, 121–140.

7

Modeling Environmental Governance in the Lake Tahoe Basin: A Multiplex Network Approach

Elizabeth A. Koebele, Stephanie Bultema, and Christopher M. Weible

Introduction

Freshwater resources, such as lakes and rivers, are susceptible to overuse and degradation due to their nature as common pool resources. To avoid these situations, governments and public agencies have created complex governance arrangements aimed at enhancing long-term water sustainability while balancing political pressures from various non-governmental forces. To understand how such arrangements manifest and function in practice, policy scholars have begun to analyze network data that represent the relational characteristics of the actors involved in water management and governance situations (Stein et al. 2011; Lienert et al.

E. A. Koebele (✉)
University of Nevada, Reno, Reno, NV, USA
e-mail: ekoebele@unr.edu

S. Bultema • C. M. Weible
University of Colorado Denver, Denver, CO, USA
e-mail: stephanie.bultema@ucdenver.edu; chris.weible@ucdenver.edu

© The Author(s) 2020
M. Fischer, K. Ingold (eds.), *Networks in Water Governance*, Palgrave Studies in Water Governance: Policy and Practice, https://doi.org/10.1007/978-3-030-46769-2_7

2013; Fliervoet et al. 2016). Network analysis can provide insight into the underlying social dynamics of water resources governance, such as patterns of coordination and conflict among various stakeholders, by illuminating how policy actors are structurally linked or separated within a particular setting. Findings from network analyses have the potential to help improve both social and environmental outcomes in natural resource management contexts (Groce et al. 2019).

While networks have been established as both a key theoretical concept and a powerful methodological tool for understanding governance arrangements (Scott and Ulibarri 2019), a number of major challenges confront their use. One of the greatest challenges is that multiple types of relationships among actors define any system of governance, but most networks only depict one face of a complex governance system. For example, examining a network of policy actors connected to one another via their shared normative beliefs about a water issue could provide one valid interpretation of a governance setting; however, a different interpretation could arise when examining the same setting through the lens of real-life interactions among actors. Each interpretation can help scholars and practitioners understand how governance happens in practice, and the relationships they depict may reinforce or otherwise affect one another in unrecognized ways.

This chapter demonstrates one way to address the challenge of capturing multiple types of relationships among actors in a governance context via the use of a type of network called a multiplex network. Multiplex networks integrate multiple layers of network data that represent different types of relationships among actors (De Domenico et al. 2013; Battiston et al. 2014). For example, a multiplex network can depict how actors share resources by integrating two networks that each represent a sharing relationship around a different type of resource (e.g. informational resources and financial resources). Multiplex networks can also be considered "multi-mode" when they include different types of nodes (Jasny 2012). For instance, a multi-mode multiplex network may show how nodes representing actors are linked to nodes representing policies they support, as well as how the actors are linked to other actors through interactions. Multiplex networks can help scholars better understand how different types of relationships influence the overall structure of a network (Gómez-Gardenes et al. 2012; Paruchuri et al. 2019; Scott and

Ulibarri 2019), and especially how one type of relationship (e.g. shared beliefs) may influence another type of relationship (e.g. coordination) (Matti and Sandström 2011). They can also provide insight into why actors may relate in one way but not in other ways (Shipilov 2012). In sum, multiplex networks can more richly reflect the reality of complex governance situations and provide a robust way to observe and analyze the various ways in which policy actors are related within them.

The empirical context for our study is the environmental governance subsystem in the Lake Tahoe Basin. Located in the western United States, Lake Tahoe is one of the world's most pristine alpine lakes and provides water and other natural resources for a host of different user groups. Over the latter half of the twentieth century through today, environmental policymaking in the Basin has focused on protecting the quality of the lake's water and its surrounding environment from negative impacts associated with local economic development. We model this subsystem during a period of major, controversial policy change—the creation and implementation of a regional plan in 2012 that sets overarching standards for both development and environmental protection efforts—using a multiplex network. To construct the network, we extract relational data from newspaper articles about the policymaking process using discourse network analysis (Leifeld 2017), a method which will be discussed in more detail below.

Our analysis is guided by the Advocacy Coalition Framework (ACF) (Jenkins-Smith et al. 2018). The ACF is one of the most established theoretical approaches for understanding public policy processes, particularly those involving natural resources. Specifically, the ACF argues that political processes can be understood through the lens of advocacy coalitions, or groups of actors who share beliefs and coordinate to achieve their policy goals. As iteration among ACF theory and empirical analysis has grown, scholars have asserted the value of using network approaches to investigate ACF concepts such as the drivers of coalition formation (Henry 2011; Ingold 2011; Ingold et al. 2017), coordination (Matti and Sandström 2011, 2013), policy change and implementation (Dela Santa 2013; Leifeld 2013b), and other aspects of policy debates surrounding controversial topics like climate change (Elgin 2015; Kukkonen et al. 2017).

We build on these studies by constructing a multiplex network that models the Lake Tahoe Basin environmental governance subsystem during a period of major policy change via two categories of relationships among actors: shared beliefs and interactions. Because the ACF assumes that multiple types of relationships drive coalition formation and influence subsystem structure, as will be described further in the next section of this chapter, using a network approach that considers various relationships among actors simultaneously is particularly appropriate for ACF applications and can provide a more holistic understanding of the governance context. We investigate three guiding research questions using this approach:

1. What similarities and differences exist in networks representing singular relationships (e.g. specific types of shared beliefs *or* interactions) among actors in the subsystem?
2. What does the integration of the singular-relationship networks into a multiplex network indicate about the overall structure of the network?
3. What can we learn about the relationships among advocacy coalitions during a period of major policy change in the Lake Tahoe Basin by examining both the singular-relationship networks and the multiplex network?

Using a multiplex network approach allows us to analyze the different types of relationships among actors both separately (RQ1) and simultaneously (RQ2), while most ACF studies examine only one type of relationship. This analysis also helps us understand the structure of a specific empirical context during an instance of major policy change (RQ3), which also provides new insight into the mechanisms driving policy change in complex water governance systems.

Theory

The ACF approaches the study of governance arrangements by focusing on the politics of the people and groups vying for influence over extended periods of time. It argues that when policy actors engage in disputes over public policy in democratic political systems, they are likely to coordinate their behavior with others who share their beliefs to achieve their policy goals (Jenkins-Smith et al. 2018; Weible et al. 2020). Actors who share beliefs and coordinate can be characterized as belonging to the same advocacy coalition. We use this basic lens to guide the collection, analysis, and interpretation of network data that captures the political dynamics among actors in the Lake Tahoe Basin environmental governance subsystem. In doing so, we utilize several definitions and theoretical insights from the ACF.

First, the ACF specifies the *policy subsystem* as the primary unit of analysis. As a subset of a broader political system, a policy subsystem focuses on a particular issue (e.g. water management) in a particular locale (e.g. Lake Tahoe Basin). Policy subsystems consist of policy actors, or the people directly or indirectly trying to influence policy decisions and outcomes. While the ACF assumes that agency lies in the individual, the force of such agency is linked to policy actors' organizational affiliation. Organizations are important because they may provide resources, help define subsystem roles, and reinforce actor identities. Because there is no readily available sampling frame from which to identify policy actors in this subsystem, our analysis includes all individuals and organizations who expressed beliefs, stated policy positions, or had recorded interactions with other policy actors in newspaper articles related to the policy change at the heart of this study.

Second, the ACF posits that policy actors are motivated by their belief systems. Central to actors' belief systems are *policy core beliefs*, which represent their overarching policy goals within a subsystem. Policy core beliefs are often described as the "glue" that bonds policy actors in forming and maintaining one or more advocacy coalitions (Weible 2005). Actors also have *secondary beliefs*, which concern their preferred means to achieve their goals. In our network, we measure policy core beliefs through actors' basic normative statements in favor of either environmental preservation or economic development within the Lake Tahoe Basin. We measure secondary beliefs through actors' statements of support or opposition for two policies. These statements can then be used to form networks that depict how actors are related via shared beliefs.

Third, another defining feature of coalitions is intentionally coordinated political activity. *Coordination* among members of a coalition can span from weak (i.e. sharing information, implicitly acting in parallel) to strong (i.e. developing and implementing joint strategic plans) (Weible and Ingold 2018). Actors may also coordinate across coalitions and are often incentivized to do so in collaborative policymaking venues (Koebele 2019). This study embraces a broad conceptualization of coordination in order to capture any intentional interactions among policy actors recorded in news media. We also document instances of unintentional or happenstance interactions, such as when actors attend the same meeting, as these circumstances may present opportunities for information-sharing, dialogue, and negotiation. Similar to how actors can be connected via their shared beliefs, coordination, whether intentional or unintentional, is another type of relationship among policy actors in a subsystem that can be depicted through a network.

In sum, the ACF argues that various types of relationships—specifically, shared policy beliefs and coordination—influence coalition formation and interaction. Multiplex networks allow for the depiction and analysis of different types of relationships both separately (i.e. through singular-relationship networks) and simultaneously (i.e. through a multiplex network that integrates multiple singular-relationship networks). This approach not only provides a way to better identify coalitions by accounting for multiple relationship types among actors, as instructed by the theory, but also helps the analyst detect places where relationships reinforce or conflict with one another. Consequently, a multiplex network can provide insight into the structure and function of a complex environmental governance subsystem, including instances of conflict and cooperation, that could not be derived from examining only one type of relationship alone.

Case: Policy Change in the Lake Tahoe Basin

Located on the border between California and Nevada, Lake Tahoe is the largest freshwater lake in the Sierra Nevada Mountains and one of the most pristine alpine lakes in the world. It is both a major tourist destination and an important water storage reservoir for the region. Conflicts between economic development interests and groups concerned with protecting the quality of the lake's pristine blue water and surrounding environment have dominated policymaking in the Basin since the mid-twentieth century. The development of roads, parking lots, homes, and other impervious features has contributed to a consistent decline in the lake's water quality and clarity by increasing erosion and reducing natural pollution filtration (Tahoe Environmental Research Center 2019). Many forests and wetlands have also been destroyed or damaged by widespread development, with significant impacts for the lake's surrounding ecosystems.

In light of these issues, California, Nevada, and the US federal government signed a Bi-State Compact that created a new governance structure for the Lake Tahoe Basin in 1969. The compact created the Tahoe Regional Planning Agency (TRPA), a bi-state agency charged with developing, monitoring, and enforcing regional management plans that consider both the Basin's economy and protection of the lake's quality and surrounding environment. During the development of the first regional plan in 1971, there was a high level of conflict: environmentalists felt the plan did not include enough environmental protection measures, while economic development groups felt the plan infringed on their right to acquire and develop land. This situation escalated into the 1980s when both sides filed lawsuits against TRPA. In response, TRPA began a multi-stakeholder, consensus-based planning process to create a new regional plan that was adopted in 1987. Since then, the Basin has embraced a collaborative approach to governance that incorporates more inclusive and participatory forms of decision-making (Imperial and Kauneckis 2003; Kauneckis and Imperial 2007). Scholars have found that collaboration among actors in the Basin has resulted in improved inter-group perceptions (Weible et al. 2011) and some degree of belief convergence among members of opposing coalitions (Weible and Sabatier 2009).

Despite the introduction of collaborative governance processes into the Basin, conflict flared once again around the next major regional plan update, which was slated to occur in 2007. Although TRPA collected and incorporated input from over 4500 stakeholders from both within and outside of the Basin, numerous political disagreements delayed the development of the plan. For example, Nevada threatened to withdraw from the Bi-State Compact in 2011 because they believed the plan would put too many restrictions on development. Following revisions to the draft plan, additional input by stakeholders and scientists, and a five-option Environmental Impact Statement assessment process, TRPA approved an updated regional plan in a 12-1 vote on December 12, 2012 (hereafter, 2012 Regional Plan). However, by February 11, 2013, some environmental groups filed a lawsuit against TRPA that sought to block implementation of the 2012 Regional Plan on the premise of insufficient environmental review. Although the lawsuit was dismissed, it signaled continuing contention among development and environmental protection interests in the Lake Tahoe Basin.

Methods

To model the Lake Tahoe Basin environmental governance subsystem during the 2012 Regional Plan process, we use discourse network analysis (Leifeld 2017). This approach combines content analysis of textual data with network analysis. Discourse—defined here as verbal interactions by and between policy actors about a given topic—is one avenue through which policy actors express preferences, engage in debates, demonstrate alliances, and persuade others. Discourse analysis can provide insight into patterns of conflict and collaboration among actors by showing the many ways in which actors connect through their stated beliefs and patterns of coordination. Discourse also draws attention to policy issues, potentially leading to shifts in agendas, policy images, and policy narratives, all of which can prompt policy change. Analyzing discourse via archived textual sources such as news media (Leifeld 2013b; Fergie et al. 2018) and congressional hearings (Fisher et al. 2013) can help to overcome issues of bias and recall error inherent in retrospective surveys or interviews of policy actors' experiences (Leifeld 2013b). This section describes our discourse network analysis approach, beginning with the collection of data via newspaper articles through a two-step coding process. We then describe the methods for constructing and analyzing our multiplex network.

Data Collection and Processing

Data were extracted from two major regional newspapers (the *Reno Gazette Journal* in Nevada and the *Sacramento Bee* in California) and from the local newspaper (the *Tahoe Daily Tribune*). We used newspaper archives and a site-specific Google search to locate and download articles published between 2005–2014 that mentioned the terms "Lake Tahoe" and "Regional Plan" (n = 175). We reviewed all articles and excluded duplicate articles, those that did not substantively mention the 2012 Plan, and those first published outside of the study period, resulting in 93 relevant articles. Most articles were drawn from the *Tahoe Daily Tribune* (n = 57), with fewer from the *Reno Gazette Journal* (n = 24) and *Sacramento Bee* (n = 12), highlighting the significant local importance of the process.

Step 1: Coding with Discourse Network Analyzer

The articles were processed using the Discourse Network Analyzer (DNA) software (Leifeld 2013a), which allows for the conversion of textual data into relational data through a manual coding process. First, a DNA database was constructed, into which media articles were uploaded. Statements (or portions of text from the articles) were then coded in relation to a priori concepts from the ACF. The final codebook consisted of three statement types:

1. Beliefs Statements connect a policy actor to a *policy core belief*, as defined by the ACF. We coded explicit, positive statements of belief in favor of environmental protection or economic development in the Basin. Statements of support for specific actions were coded when they clearly indicated the actor's desire to protect the environment or develop the area. We did not code statements of fact, neutrality, or negative belief because these would require making assumptions about the beliefs that actors hold.
2. Interaction Statements connect two actors via an intentional (deliberate) interaction or an unintentional (happenstance) interaction. Intentional interactions represent *coordination* among actors, as

defined by the ACF, while unintentional interactions do not explicitly line up with ACF concept but may represent a weak form of coordination, particularly in collaborative processes that engage diverse actors. We incorporate both forms of interactions in the analysis below.

3. Policy Position Statements connect an actor to either the 2012 Regional Plan or the 1969 Bi-State Compact via an expression of support or opposition. Policy positions represent an actor's *secondary beliefs*, as defined by the ACF.

We developed and refined the coding framework through three rounds of intercoder reliability testing.[1] All remaining articles were coded by one coder and reviewed by a second coder. Cases of ambiguity were discussed among three coders until agreement could be reached.

In this process, the article was considered the unit of analysis. Thus, if the same relationship occurred more than once within an article, it was only coded in the first instance. This resulted in a total of 606 coded statements by 72 unique actors. We coded both direct quotes and narrative statements by the article authors. When an individual's organizational affiliation was stated in the article, both the individual and their organization were recorded. When statements were recorded for multiple individuals representing a single organization, the individual reports were aggregated to the organization level. Most of the policy actors in our network represent organizations (82%, $n = 59$); policy actors are only represented as individuals if the statement was not made on behalf of an affiliated organization. The final dataset reflects the publicly stated beliefs, interactions, and policy positions of individuals and organizations involved in Lake Tahoe Basin environmental governance during the 2012 Regional Plan process.

---

[1] In Rounds 1 and 2, ten randomly selected articles were coded independently by three coders. Codes were compared and the codebook was revised to clarify directions and coding options. The process was repeated in Round 3, where three coders came to full agreement on codes across the same ten articles coded in Round 2.

Step 2: Coding Coalition Membership

Each policy actor was then assigned to either a pro-development or a pro-environment coalition based on their policy core beliefs and intentional interactions (i.e. coordination), as suggested by the ACF. These two coalitions were detected in a coalitional analysis previously conducted in the Lake Tahoe Basin (Weible and Sabatier 2009). One coder assigned numeric values to each pro-environment (+1) and pro-development (−1) belief in the dataset. These values were then summed for each actor and normalized to their total number of belief statements. Additionally, each time an actor intentionally interacted with a group/actor that had a positive belief score, the interaction was coded as +1, whereas an interaction with an actor who had a negative belief score was coded as −1. Interaction values were summed for each actor and normalized to their total number of intentional interaction statements. The normalized belief and interaction scores were then summed to form a "coalition affiliation score" (CA score), which ranges from −2 (strong pro-development affiliation) to +2 (strong pro-environment affiliation).

Because an actor's CA score may fall anywhere in this range, a CA score indicates both coalition membership and the strength of an actor's coalition affiliation. Scores of zero indicate that no coalition affiliation was evident (i.e. actors only had policy positions or unintentional interactions) or that we recorded equal numbers of pro-development and pro-environment sentiments. This approach for identifying coalitions is somewhat unique in that actors may be grouped into coalitions due only to their interactions, even if they fail to state clear policy core beliefs. While it may potentially over-emphasize the importance of interactions as a conduit for coordinated political activity, it allows us to capture both core members of coalitions and those that are more auxiliary in our analysis of the subsystem.

Network Data Analysis

Relational data were exported from DNA for quantitative network analysis in Gephi 9.2. First, we created a singular-relationship network for each statement type using separate node and edges tables. The beliefs network is a two-mode network that connects actors (nodes) to beliefs (nodes) via belief statements (edges). The interactions network is a one-mode network that connects actors (nodes) to other actors via interaction statements (edges). The policy positions network is a two-mode network that connects actors (nodes) to policy positions (nodes) via policy position statements (edges). In this third network, two nodes were created for each policy to differentiate those connected an actor to a policy via support or via opposition. Additional attribute data such as node type, coalition affiliation, edge type (e.g. intentional vs. unintentional interaction), and node/edge color were added to these three singular-relationship networks, where applicable, to assist with network analysis and visualization. Next, we combined these three singular-relationship network files to produce a fourth network: our multi-mode, multiplex network. These four networks were then assessed using descriptive network analysis. All networks were treated as undirected networks.

Results

Figure 7.1 displays the four networks (three singular-relationship networks and the combined multiplex network), with each actor labeled as belonging to an advocacy coalition (pro-environment, pro-development, or neutral) based on their CA score. In each network, light green numbered squares represent members of the pro-environment coalition and light blue numbered triangles represent members of the pro-development coalition. Non-affiliated actors are indicated by light grey circles. The other unique node and edge variations in singular-relationship networks are depicted via the Node Key and Edges Key shown below the network maps. Each node is sized relative to its prominence in the network based on the number of statements associated with the node. Table 7.1 provides an overview of basic network statistics for each network, as well as definitions of these statistics. The Appendix provides a list linking node numbers to actor names and characteristics.

Of the three singular-relationship networks (beliefs, interactions, policy positions), the interaction network is the largest, with 263 interaction

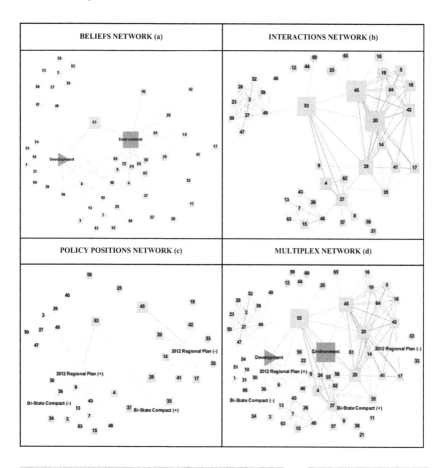

Fig. 7.1 Disaggregated multiplex network

Table 7.1 Overview of network statistics

	Statements (edges)[a]	Nodes[b]	Average degree[c]	Modularity[d]	Communities[e]
Beliefs network	219	51	8.59	0.18	2
Interactions network	263	44	11.96	0.47	6
Policy positions network	124	38	6.53	0.5	3
Multiplex network	606	72	16.83	0.37	4

[a]Edge: an edge is drawn as a line connecting two nodes, which represents a relationship between those nodes
[b]Node: a node (or vertex) is drawn as a shape (typically a circle) representing the entities that make up a network
[c]Average degree: measures the average number of connections reported for all nodes in a network
[d]Modularity: measures the strength of division of a network into modules (also called groups, clusters, or communities). Networks with high modularity (closer to 1.0) have dense connections between the nodes within modules, but sparse connections between nodes in different modules
[e]Communities: the modules detected by a community detection algorithm (such as modularity). Nodes in a community have stronger connections with nodes in their community than with nodes in other communities. A network with fewer communities indicates a more cohesive network, while a network with many communities is indicative of a more divided network

statements among 44 actors. The beliefs network is also sizable, with 219 belief statements recorded for 51 actors. The smallest network, the policy positions network, includes 124 policy position statements made by 38 actors. To answer the questions posed above, the remainder of this section will briefly describe key insights from each statement network, as well as from the multiplex network as a whole.

Beliefs Network

The beliefs network (Fig. 7.1a) depicts each actor who expressed either a pro-environment or pro-development belief in our dataset, which we use to measure the ACF concept of policy core beliefs. The environment belief node has a degree of 140, while the development belief node has a degree of 79 (Appendix Table 7.2). Because degree measures the total number of connections reported for a node, this indicates that actors expressed pro-environment beliefs almost twice as frequently as pro-development beliefs. Of the 49 actor nodes in this network, 18 expressed only pro-environment beliefs, 17 expressed only pro-development beliefs, and 14 expressed a combination of beliefs. The actor with the most belief statements was TRPA (node 53), the agency leading the Regional Plan process. TRPA expressed both types of beliefs, but tended to express pro-environment beliefs more often, contributing to their moderately positive CA score (0.4) and consequent categorization as a member of the pro-environment coalition.

Of the 18 actors with mixed belief statements, 10 were assigned to the pro-development coalition based on their CA score, while 7 were assigned to the pro-environment coalition (including TRPA), and 1 was unaffiliated. This indicates that about 37% of actors engaged in discourse that recognized the beliefs held by their opponents. For example, during a highly contentious moment in which Nevada (a pro-development coalition member) threatened to exit the Bi-State Compact because they feared the 2012 Regional Plan would be overly restrictive of development, Nevada Senator John Lee stated, "We just want to protect our property rights without harming the lake … We understand it's a jewel. If we harm that jewel, we harm everything we have in Nevada."[2] Although this statement captures the Nevada Senate's desire to protect the state's property rights first and foremost, Senator Lee "tips his hat" to the policy core beliefs of the opposing coalition (i.e. those focused primarily on environmental protection). Given the importance of Tahoe's unique natural environment to both coalitions' goals, this kind of language is somewhat unsurprising. Moreover, the collaboratively-oriented process used to develop the Regional Plan encouraged policy actors to dialogue and seek points of consensus, which may have further contributed to such mixed expressions of belief.

---

[2] DeLong, J. 2011. "TRPA: Nevada bill could help force meaningful change." *Reno Gazette-Journal.*

Interaction Network

The interactions network (Fig. 7.1b) depicts both intentional and unintentional interactions among actors. From this network, it is obvious that the majority of interactions were deemed intentional, which we use to measure the ACF concept of coordination. For example, Friends of the West Shore (node 20) and the Sierra Club (node 45) jointly sued TRPA (node 53) following the passage of the 2012 Regional Plan, while a number of business-focused groups, such as the South Shore Chamber of Commerce (node 49) and the Incline Village Board of Realtors (node 23), jointly signed and submitted a formal statement in support of the Regional Plan to counter the suit. Both of these examples represent intentional interactions (i.e. coordination).

Moreover, as is expected by the ACF, the majority of intentional interactions happened between actors from the same coalition (153 acts of coordination among pro-environment actors; 47 acts of coordination among pro-development actors). Only 11 of the 44 actors in this network who were affiliated with a coalition engaged in cross-coalition coordination (12 acts), meaning they intentionally interacted with a member of the opposing coalition.[3] This suggests that even though the processes through which the Regional Plan was developed were intended to be somewhat collaborative, actors usually coordinated with others who were affiliated with the same coalition and less frequently coordinated with members of the opposing coalition.

Policy Positions Network

The policy positions network (Fig. 7.1c) depicts each actor who expressed either a supportive or oppositional position toward either the 2012 Regional Plan or the 1969 Bi-State Compact. These statements represent

---

[3] A supplemental analysis suggests that, in general, those with relatively low coalition affiliation scores are more likely to engage in interactions across coalitions. Such a pattern contributes to accumulating evidence that extreme beliefs are associated with more polarized positions within coalitions (e.g. see Elgin 2015; Weible 2005). However, empirically testing and elaborating on such tendencies in relation to existing literature is beyond the scope of this chapter.

the ACF concept of secondary beliefs, or the mechanisms through which actors want to achieve their policy core beliefs. Each policy has supporters from at least two of the three major groups in the network (the pro-environment coalition, the pro-development coalition, or unaffiliated). Both support for and opposition to the Bi-State Compact was expressed by members of each major coalition, suggesting that the policy did not attract uniform support or opposition by one side.

The most discussed policy position was support for the 2012 Regional Plan, which was expressed at least once by 14 pro-development actors, 7 pro-environment actors, and 1 unaffiliated actor. For example, in the statement filed by the business groups in response to the lawsuit described above, they argued that the 2012 Regional Plan was necessary for the Basin to flourish, environmentally and economically: "'No one is clamoring for more development. Rather, environmental redevelopment of the basin's built environment [as advocated for in the Plan] is necessary to accelerate achievement of TRPA's (environmental) thresholds, improve the basin's economy and restore the basin's communities.'"[4] In a similar vein, a representative of the League to Save Lake Tahoe, an environmental group, argued that it "'is not a perfect plan, but … has the potential to help Tahoe's environment through multiple safeguards that require restoration and environmental improvements with any new development or redevelopment.'"[5] Importantly, these statements supporting the 2012 Regional Plan also echo the idea discussed above of coalitions "tipping the hat" to one another's policy core beliefs in a collaborative policymaking context.

No pro-development actors expressed opposition to the 2012 Regional Plan, meaning this policy position was only held by pro-environment ($n = 10$) and neutral ($n = 2$) actors. For example, a representative of the Sierra Club, a major environmental group involved in the lawsuit against TRPA after the Regional Plan's adoption, framed the plan as a loss for environmental protection interests by saying, "'This is a wrenching departure from past practice and is not in line with the spirit or law of the

---

[4] Lotshaw, T. 2013. "TRPA, local governments, other groups defend regional plan in court." *Tahoe Daily Tribune.*

[5] DeLong, J. 2014. "TRPA wins Tahoe lawsuit." *Reno Gazette-Journal.*

bi-state compact created to protect the lake.'"[6] However, as indicated above, many members of the pro-environment coalition also expressed support for the 2012 Regional Plan, signifying a rift in secondary beliefs among members of the pro-environment coalition.

Multiplex Network

Each singular-relationship network described above (beliefs, interactions, policy positions) has different structural characteristics and provides unique insight that can only be understood by assessing it individually. However, integration of these networks into a multiplex network provides another, more holistic view of the overall network structure. This is because the multiplex network depicts how relationships may be reinforced through multiple types of connections among actors. For example, two actors may be connected by similar beliefs, policy positions, and joint interactions, which may indicate a very strong relationship between the actors. Weaker relationships may be evident through only one relationship type, such as when two actors have interacted but do not share policy core beliefs or policy positions. In this way, the multiplex network not only allows a more theoretically-informed analysis of coalitions per the ACF, but it also offers a more encompassing picture of the relationships among policy actors who participated in subsystem discourse about the 2012 Regional Plan than does any singular-relationship network.

The multiplex network (Fig. 7.1d) depicts how all actors in the network are connected via all types of relationship coded for in the discourse network analysis: shared beliefs, policy positions, and intentional and unintentional interactions. The multiplex network weighs all of these connections equally, so, for example, an intentional interaction among actors is not depicted or interpreted as more important than an unintentional interaction when calculating statistics for the network. The multiplex network includes 24 pro-development coalition members, 33 pro-environment coalition members, 9 unaffiliated actors, and 6 non-actor nodes (2 beliefs and 4 policy positions). The average coalition

---

[6] DeLong, J. 2013. "Critics sue to block new land use plan for Lake Tahoe." *Reno Gazette-Journal.*

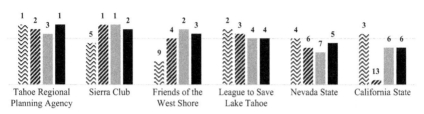

Fig. 7.2 Relative rank of major policy actors across networks

affiliation scores are −0.66 for the pro-development coalition and 0.68 for the pro-environment coalition, suggesting that the strength of pro-development or pro-environment sentiment in the two coalitions' discourse did not differ greatly.

One way to examine similarities and differences across the four networks—and to understand what the multiplex network adds to our interpretation of the governance system—is to examine the relative prominence of major policy actors (i.e. those who were associated with the most statements) for each of the four networks.[7] This analysis provides some insight into which policy actors may play key roles within the subsystem. To calculate this, actors were ranked by their degree in the full multiplex network, with the top-ranked actors having the highest degree. As can be seen in Fig. 7.2, the top six actors in the multiplex network include the following: (1) Tahoe Regional Planning Agency, (2) Sierra Club, (3) Friends of the West Shore, (4) League to Save Lake Tahoe, (5) Nevada State, and (6) California State. The rank of each of these actors is also shown for each network individually in Fig. 7.2.

---

[7] We must add a caveat to this analysis in that the centrality measures across our networks are not directly comparable, given that some networks are one-mode while others are two-mode. We circumvent this concern by comparing the relative ranking of each actor's centrality (versus raw degree) within each network.

Some key variations across networks can be gleaned from Fig. 7.2. First, TRPA is one of the top three actors across all networks and the top-ranked actor in both the beliefs and multiplex networks, which reinforces their prominence in the conversation about the 2012 Regional Plan. Other actors' prominence varies across networks, however. For instance, Friends of the West Shore, a pro-environment coalition member, was ranked relatively low in the beliefs network, potentially signifying that they were only a minor actor in the process. However, they are ranked higher in the other singular-relationship networks and third overall in the multiplex network, which is perhaps a better reflection of their influence in subsystem discourse: Friends of the West Shore was one of the groups that coordinated to oppose to the 2012 Regional Plan and to eventually sue TRPA after the Regional Plan's enactment. Similarly, California State was ranked low (13th) in the policy positions network, but was the 6th most prominent actor in the multiplex network. In this way, Fig. 7.2 exemplifies the potential pitfalls of identifying key actors in a governance process based on a network representing only one type of relationship.

Discussion and Conclusion

This chapter models the Lake Tahoe Basin environmental governance subsystem during a period of policy change using discourse network analysis of newspaper articles, which were used to construct a multiplex network. One of our key assumptions was that a policy subsystem is multifaceted and no single network type can capture the different faces of its complexity. This assumption is a foundational tenet of the ACF, which was used to guide this analysis. We argue that using a network approach that incorporates multiple types of relationships simultaneously allows for a more precise application of the ACF as well as a more holistic look at the inner workings of a complex governance system. Moreover, to the extent that one data source (newspapers) can be used to understand a complex water governance system, this chapter integrates three different types of singular-relationship networks (beliefs, policy positions, and interactions) into a multiplex network. As part of constructing the networks, we developed a new method for identifying coalitions (the CA score), which contributes an analytical innovation to the ACF literature that may be particularly helpful for scholars wishing to detect coalitions via media and discourse analysis techniques.

We gain a number of unique insights into the Lake Tahoe Basin environmental governance subsystem between 2005 and 2014 through this analysis. By analyzing networks based on singular relationships (RQ1), we detected the presence of fairly distinct pro-environmental and pro-development coalitions and found that most coordination occurred among members of the same coalition. Despite this, members of both coalitions exhibited "tipping of the hat" discourse in which they publicly recognized the central policy core beliefs of their opposition, even if they did not intentionally work together. Actors also exhibited some concurrence on the 2012 Regional Plan, and both support for and opposition to it was present within the pro-environment coalition. This suggests a rift within the pro-environment coalition over secondary beliefs: coalition members agreed that conserving the environment is a priority, but they disagreed about whether the 2012 Regional Plan was the best way to achieve that. These findings highlight the complexity of environmental governance in the Lake Tahoe Basin, wherein there is a strong interdependence between the preservation of a healthy lake environment and continued economic growth and development.

When combining the beliefs, interactions, and policy positions networks into a multiplex network, we provide a holistic image of the complex network formed by the different types of relationships among actors (RQ2). The most prominent actors in the multiplex network were not necessarily frequent participants in all types of discourse, which suggests that looking at a singular-relationship network on its own may present an incomplete picture of the subsystem. In this regard, the multiplex network provides a more comprehensive model of the subsystem, particularly when enriched by additional analyses of and comparisons with the networks that were integrated to form it.

Our findings suggest a number of practical lessons about governance in the Lake Tahoe Basin and the coalitions of actors that participated in it during a period of major policy change (RQ3). First, members of both major coalitions were publicly willing to recognize the value of preserving the environment, whether to promote its inherent value or to bolster the economy of the region. While contentiousness among the two coalitions is likely to endure into the future due to deeply-held policy core beliefs, starting conversations about environment management in the Basin from this shared value could help promote more effective collaboration and consensus-building. A second lesson involves the extent to which coalitions overlap and sometimes splinter. While we are far from identifying the situations when this might happen, it is clear from this study that members of a coalition might hold very different beliefs about what policy option can best achieve their goals. How long such a division might last, or what factors may reengage the full coalition, are empirical questions worth exploring. Similarly, members from opposing coalitions may have different policy core beliefs but may see promise in a single, collaboratively-developed policy solution, like the 2012 Regional Plan. This may be because they come to recognize, through engaging in the collaborative policy process, that they have common or interdependent goals; or, they may believe that a single policy instrument is broad enough to help them obtain their own goals simultaneously and without direct conflict.

There are also a number of limitations to this study that point to directions for future research. Although we constructed our networks based on 606 unique statements, they were drawn from a fairly limited number of media articles ($n = 93$), which were mainly published in one source. Consequently, our networks are limited to the discourse that was selectively included or created by journalists and may therefore reflect journalists' style and focus. While media data are often easier to obtain (though the process of accessing historical articles is becoming increasingly difficult with small newspapers that have fewer resources to maintain archives) and less intensive to process (because it is not entangled with other policy process documents, as it would be in meeting minutes), it is inherently more simplistic than the kind of discourse data that may be obtained from other sources. This may also cause scholars to overlook important coalition dynamics that fail to grab media attention, while over-emphasizing those that do.

To overcome these limitations, scholars should collect discourse data from multiple sources and compare patterns across sources. Relatedly, while the multiplex network approach used here provides a more robust understanding of the subsystem discourse than any of the singular-relationship networks, the largely quantitative nature of the output of the DNA program limits the ability of scholars to understand the mechanisms driving network structure. In the future, scholars can address this problem by using network analysis as one method in broader, mixed-method studies, wherein multiple sources of information (interviews, qualitative analysis of textual data) can be triangulated to enhance the researcher's understanding of *why* networks are structured in the observed ways. Moreover, disaggregating the networks into the periods before and after the 2012 Regional Plan was enacted could provide insight into how policy change affects network structure.

The extent to which the insights and lessons gleaned from this analysis are generalizable to other systems is an open question. The Lake Tahoe Basin is, in many ways, different than many water governance settings. It straddles two states in the United States, has a long history of adversarial and, more recently, collaborative approaches to governance, and has a centralized governance structure with a single regional entity charged with regulating and enforcing public policies in managing the Basin environment. While this context makes the Lake Tahoe Basin subsystem unique, it also shares characteristics with other complex water governance systems: there are coalitions, dilemmas in balancing economic and environmental priorities, a complex role for science in decision-making, a shift toward collaborative practices, and repeated interactions among policy actors that can, at times, span decades. The next step is to apply the methodological techniques used here to other systems. If done, we will have the opportunity to make both methodological and theoretical contributions to the study of complex governing arrangements of water resources.

Acknowledgments Thank you to Ashlyn Maher, who aided in data collection and coding. This chapter was presented at the 2019 Midwest Political Science Association Annual Conference.

Appendix

Node labels and characteristics

Node	Node ID	Node type	Degree[a]	Betweenness centrality[b]	Coalition affiliation score
2012 Regional Plan (−)	2012 Regional Plan (−)	Policy	54	144.44	
2012 Regional Plan (+)	2012 Regional Plan (+)	Policy	39	249.45	
Alex Mourelatos	1	Actor	1	0.00	−0.5
Barton Health	2	Actor	10	1.08	−1
Bi-State Compact (−)	Bi-State Compact (−)	Policy	13	76.01	
Bi-State Compact (+)	Bi-State Compact (+)	Policy	18	4.15	
California Senate	3	Actor	4	0.93	0.6
California State	4	Actor	37	20.09	0.3
California Watershed Network	5	Actor	15	0.00	0.3
Carl Ribaudo	6	Actor	7	2.27	−0.1
Carson City	7	Actor	8	3.54	−0.2
Center for Collaborative Policy at California State University Sacramento	8	Actor	4	0.00	0.5
Clarence E. Heller Foundation	9	Actor	3	0.05	0.5
Dennis Crab	10	Actor	1	0.00	−0.5
Development	Development	Belief	79	767.99	
Dianne Feinstein	11	Actor	1	0.00	0.5
District 5	12	Actor	1	0.00	0
Douglas County	13	Actor	9	0.14	−1.2
Earthjustice	14	Actor	15	2.02	1
El Dorado County	15	Actor	12	8.65	0.4
Environment	Environment	Belief	140	853.00	
Friends of Burke Creek	16	Actor	8	0.00	0.3
Friends of Lake Tahoe	17	Actor	21	1.62	1

(continued)

(continued)

Node	Node ID	Node type	Degree[a]	Betweenness centrality[b]	Coalition affiliation score
Friends of Tahoe Vista	18	Actor	16	0.34	0.3
Friends of the Crystal Bay/ Brockway	19	Actor	15	0.00	0.3
Friends of the West Shore	20	Actor	67	54.68	1.7
Green Party	21	Actor	1	0.00	0
Greta Hambsch	22	Actor	1	0.00	0.5
Incline Village Board of Realtors	23	Actor	9	0.00	−1
Jim Baetge	24	Actor	1	0.00	0.5
John A. Mendez	25	Actor	10	0.25	0
Lahontan Regional Water Quality Control Board	26	Actor	5	0.06	0.5
Lake Tahoe Community College	27	Actor	10	1.08	−1
Lake Tahoe Visitors Authority	28	Actor	9	0.00	−1
League to Save Lake Tahoe	29	Actor	65	415.23	1.2
Lew Feldman	30	Actor	2	0.05	−0.5
Lou Pierini	31	Actor	1	0.00	−0.5
Mara Bresnick	32	Actor	1	0.00	0
Margaret Martini	33	Actor	1	0.00	0
Nevada Assembly	34	Actor	1	0.00	0
Nevada Conservation League	35	Actor	17	13.92	1.5
Nevada Senate	36	Actor	14	13.21	−0.6
Nevada State	37	Actor	39	80.34	0.6
Norma Santiago	38	Actor	4	70.09	0
North Lake Tahoe Chamber of Commerce	39	Actor	10	1.08	−1
North Lake Tahoe Resort Association	40	Actor	2	0.00	−0.5

(continued)

(continued)

Node	Node ID	Node type	Degree[a]	Betweenness centrality[b]	Coalition affiliation score
North Tahoe Citizen Action Alliance	41	Actor	18	0.81	0.9
North Tahoe Preservation Alliance	42	Actor	35	14.55	1.2
Placer County	43	Actor	12	9.56	−0.3
Reno local government	44	Actor	1	0.00	0
Sierra Club	45	Actor	87	138.56	1.8
Sierra Nevada Alliance	46	Actor	1	0.00	0.5
Sierra Nevada Association of Realtors	47	Actor	11	16.51	−0.9
South Lake Tahoe City	48	Actor	22	27.55	−0.3
South Shore Chamber of Commerce	49	Actor	17	125.47	−0.5
South Tahoe Alliance of Resorts	50	Actor	10	1.08	−1
Stephen Reinhard	51	Actor	1	0.00	−0.5
Tahoe Douglas Visitors Authority	52	Actor	9	0.00	−1
Tahoe Regional Planning Agency	53	Actor	139	547.09	0.4
Tahoe Sierra Board of Realtors	54	Actor	1	0.00	−0.5
The Pathway Forum	55	Actor	1	0.00	0.5
University of California Davis	56	Actor	3	0.00	0.5
US Bureau of Land Management	57	Actor	4	0.00	1
US Department of the Interior	58	Actor	1	0.00	0.5
US District Court	59	Actor	1	0.00	0.5

(continued)

(continued)

Node	Node ID	Node type	Degree[a]	Betweenness centrality[b]	Coalition affiliation score
US Environmental Protection Agency	60	Actor	1	0.00	0
US Fish and Wildlife Service	61	Actor	1	0.00	0.5
USDA Forest Service	62	Actor	11	1.01	0.9
Washoe County	63	Actor	7	0.05	−0.7
Washoe Meadows Community	64	Actor	15	0.00	0.3
Washoe Tribe	65	Actor	1	0.00	0.5
Workforce Housing Association	66	Actor	1	0.00	−0.5

[a]Degree: the total number of connections of all types reported for a single node. Degree is calculated here based on the multiplex network
[b]Betweenness centrality: the frequency with which a node appears on the shortest path between any given nodes in the network. A higher betweenness centrality score means the node is more central in the network. Betweenness centrality is calculated here based on the multiplex network

References

Battiston, F., Nicosia, V., & Latora, V. (2014). Structural Measures for Multiplex Networks. *Physical Review. E, Statistical, Nonlinear, and Soft Matter Physics, 89*(3), 032804.

De Domenico, M., Solé-Ribalta, A., Cozzo, E., Kivelä, M., Moreno, Y., Porter, M. A., & Arenas, A. (2013). Mathematical Formulation of Multilayer Networks. *Physical Review X, 3*(4), 041022.

Dela Santa, E. (2013). The Politics of Implementing Philippine Tourism Policy: A Policy Network and Advocacy Coalition Framework Approach. *Asia Pacific Journal of Tourism Research, 18*(8), 913–933.

Elgin, D. J. (2015). Utilizing Hyperlink Network Analysis to Examine Climate Change Supporters and Opponents. *Review of Policy Research, 32*(2), 226–245.

Fergie, G., Leifeld, P., Hawkins, B., & Hilton, S. (2018). Mapping Discourse Coalitions in the Minimum Unit Pricing for Alcohol Debate: A Discourse Network Analysis of UK Newspaper Coverage. *Addiction, 114*(4), 741–753.

Fisher, D. R., Waggle, J., & Leifeld, P. (2013). Where Does Political Polarization Come From? Locating Polarization Within the US Climate Change Debate. *American Behavioral Scientist, 57*(1), 70–92.

Fliervoet, J., Geerling, G., Mostert, E., & Smits, A. (2016). Analyzing Collaborative Governance Through Social Network Analysis: A Case Study of River Management Along the Waal River in The Netherlands. *Environmental Management, 57*(2), 355–367.

Gómez-Gardenes, J., Reinares, I., Arenas, A., & Floría, L. M. (2012). Evolution of Cooperation in Multiplex Networks. *Scientific Reports, 2,* 620.

Groce, J. E., Farrelly, M. A., Jorgensen, B. S., & Cook, C. N. (2019). Using Social-Network Research to Improve Outcomes in Natural Resource Management. *Conservation Biology, 33*(1), 53–65.

Henry, A. D. (2011). Ideology, Power, and the Structure of Policy Networks. *Policy Studies Journal, 39*(3), 361–383.

Imperial, M. T., & Kauneckis, D. (2003). Moving from Conflict to Collaboration: Watershed Governance in Lake Tahoe. *Natural Resources Journal, 43*(4), 1009–1055.

Ingold, K. (2011). Network Structures Within Policy Processes: Coalitions, Power, and Brokerage in Swiss Climate Policy. *Policy Studies Journal, 39*(3), 435–459.

Ingold, K., Fischer, M., & Cairney, P. (2017). Drivers for Policy Agreement in Nascent Subsystems: An Application of the Advocacy Coalition Framework to Fracking Policy in Switzerland and the UK. *Policy Studies Journal, 45*(3), 442–463.

Jasny, L. (2012). Baseline Models for Two-Mode Social Network Data. *Policy Studies Journal, 40*(3), 458–491.

Jenkins-Smith, H. C., Nohrstedt, D., Weible, C. M., & Ingold, K. (2018). The Advocacy Coalition Framework: An Overview of the Research Program. In C. M. Weible & P. A. Sabatier (Eds.), *Theories of the Policy Process* (4th ed., pp. 135–172). New York, NY: Routledge.

Kauneckis, D., & Imperial, M. T. (2007). Collaborative Watershed Governance in Lake Tahoe: An Institutional Analysis. *International Journal of Organization Theory and Behavior, 10*(4), 503–546.

Koebele, E. (2019). Cross-Coalition Coordination in Collaborative Environmental Governance Processes. *Policy Studies Journal.* https://doi.org/10.1111/psj.12306.

Kukkonen, A., Ylä-Anttila, T., & Broadbent, J. (2017). Advocacy Coalitions, Beliefs and Climate Change Policy in the United States. *Public Administration, 95*(3), 713–729.

Leifeld, P. (2013a). *Discourse Network Analyzer Manual*. Retrieved from http://www.philipleifeld.de.

Leifeld, P. (2013b). Reconceptualizing Major Policy Change in the Advocacy Coalition Framework: A Discourse Network Analysis of German Pension Politics. *Policy Studies Journal, 41*(1), 169–198.

Leifeld, P. (2017). Discourse Network Analysis: Policy Debates as Dynamic Networks. In J. N. Victor, M. N. Lubell, & A. H. Montgomery (Eds.), *The Oxford Handbook of Political Networks* (pp. 301–326). Oxford: Oxford University Press.

Lienert, J., Schnetzer, F., & Ingold, K. (2013). Stakeholder Analysis Combined with Social Network Analysis Provides Fine-Grained Insights Into Water Infrastructure Planning Processes. *Journal of Environmental Management, 125*, 134–148.

Matti, S., & Sandström, A. (2011). The Rationale Determining Advocacy Coalitions: Examining Coordination Networks and Corresponding Beliefs. *Policy Studies Journal, 39*(3), 385–410.

Matti, S., & Sandström, A. (2013). The Defining Elements of Advocacy Coalitions: Continuing the Search for Explanations for Coordination and Coalition Structures. *Review of Policy Research, 30*(2), 240–257.

Paruchuri, S., Goossen, M. C., & Phelps, C. C. (2019). Conceptual Foundations of Multilevel Social Networks. In S. E. Humphrey & J. M. LeBreton (Eds.), *The Handbook of Multilevel Theory, Measurement, and Analysis*. Washington, DC: American Psychological Association.

Scott, T. A., & Ulibarri, N. (2019). Taking Network Analysis Seriously: Methodological Improvements for Governance Network Scholarship. *Perspectives on Public Management and Governance, 2*(2), 89–101.

Shipilov, A. (2012). Strategic Multiplexity. *Strategic Organization, 10*(3), 215–222.

Stein, C., Ernstson, H., & Barron, J. (2011). A Social Network Approach to Analyzing Water Governance: The Case of the Mkindo Catchment, Tanzania. *Physics and Chemistry of the Earth, Parts A/B/C, 36*(14–5), 1085–1092.

Tahoe Environmental Research Center. (2019). *Tahoe: State of the Lake Report 2019*. Davis, CA: University of California Davis.

Weible, C. M. (2005). Beliefs and Perceived Influence in a Natural Resource Conflict: An Advocacy Coalition Approach to Policy Networks. *Political Research Quarterly, 58*(3), 461–475.

Weible, C. M., & Ingold, K. (2018). Why Advocacy Coalitions Matter and Practical Insights About Them. *Policy and Politics, 46*(2), 325–343.

Weible, C. M., & Sabatier, P. A. (2009). Coalitions, Science, and Belief Change: Comparing Adversarial and Collaborative Policy Subsystems. *Policy Studies Journal, 37*(2), 195–212.

Weible, C. M., Siddiki, S. N., & Pierce, J. J. (2011). Foes to Friends: Changing Contexts and Changing Intergroup Perceptions. *Journal of Comparative Policy Analysis: Research and Practice, 13*(5), 499–525.

Weible, C. M., Ingold, K., Nohrstedt, D., Henry, A. D., & Jenkins-Smith, H. (2020). Sharpening Advocacy Coalitions. *Policy Studies Journal.* https://doi.org/10.1111/psj.12360.

8

Collaboration in Water Quality Management: Differences in Micro-Pollutant Management Along the River Rhine

Laura Mae Jacqueline Herzog and Karin Ingold

Introduction

Collaboration is frequently seen as one of the key aspects of managing user conflicts or complex problems (Ansell and Gash 2008; Crona and Bodin 2009). This positive association with collaboration and its consequences is reasonable if one defines it as "working together to produce a 'beneficial' outcome to all parties involved" (see Huxham 1993, p. 603; O'Leary and Vij 2012, p. 510). In reality, complex issues such as water scarcity, environmental pollution, or climate change are not easily

L. M. J. Herzog (✉)
University of Osnabrück, Osnabrück, Germany
e-mail: laura.herzog@uni-osnabrueck.de

K. Ingold
Federal Institute of Aquatic Science and Technology, Eawag, Dübendorf, Switzerland

University of Bern, Bern, Switzerland
e-mail: karin.ingold@ipw.unibe.ch

© The Author(s) 2020
M. Fischer, K. Ingold (eds.), *Networks in Water Governance*, Palgrave Studies in Water Governance: Policy and Practice, https://doi.org/10.1007/978-3-030-46769-2_8

managed. Furthermore, the quality of actor collaboration in such contexts of complex problems and its outcomes depend, to a large extent, on different drivers and context factors (Feiock and Scholz 2010; Calanni et al. 2015; Berardo and Lubell 2016). With a few exceptions (see Lubell et al. 2012), the literature either focuses on factors that impact collaboration and its quality (Ingold and Fischer 2014), or on outcomes of collaboration (Ostrom 1998). We apply an integrative approach and consider both: Our study assesses whether the intensity and structure of actor collaboration in the context of an environmental policy problem can provide insights on results of collaborations, that is, differences in policy outcomes.

We do so borrowing from relevant literature in policy studies and relying on the logic of the social-ecological system framework (SES). The SES framework, developed by Ostrom (2005), focuses on problems related to natural and common pool resources (CPR). The framework focuses on actors' interaction that happens in a so-called action situation and produces outcomes as a means to solve the environmental problem under study (through, for instance, new management rules). The framework is a tool with which one can study the interplay of the social system and the ecological system in order to identify causes, extent and quality of environmental problems, as well as potential solutions and their impact on the ecological and social systems. The action situation is located in between the broader resource and governance contexts that set its conditions (McGinnis and Ostrom 2014). The action situation describes the locus of actor interaction, where actors are concerned with the given environmental issue. In the present study, we examine actor collaboration in the context of water quality management, which we conceptualize as an actor collaboration network that is situated in the action situation of the SES under study. In the chapter's conclusion, we discuss the outcome of such a collaboration, that is, the policies tackling the water quality problem under study.

To study actor collaboration in the context of an environmental problem, we focus on a specific type of a water quality issue: micro-pollutants in the Rhine catchment area. Micro-pollutants in surface water represent a rising concern, as a variety of sources and entry paths into surface water exist. Micro-pollutants stem from pesticide use in agriculture, biocides applied to buildings, as well as from a variety of pharmaceuticals. They can have adverse effects on aquatic ecosystems and impacts on human health. The latter is specifically an issue in regions where surface water is used for production of drinking water which could contain persistent micro-pollutants. Furthermore, when not filtered out of the water, micro-pollutants persist in water bodies, crossing different competent jurisdictions, and thus holding administrative bodies from the regional to the international level accountable for their management. Micro-pollutants are a typical multi-level and cross-sectoral policy problem of the sort that requires that actors from several sectors, various administrative levels, and different jurisdictions collaborate in deciding upon and implementing appropriate public policies.

We are interested in how actors in different sub-catchments along the Rhine tackle this water pollution issue. The SES framework provides us with an ideal analytical structure, as it helps to conceptualize the problem (through the impact of human activities on the resource water), and to investigate the action situation of actors collaborating to find a solution. We are particularly interested in policymaking: our study's action situation comprises all public and private collective actors directly or indirectly deciding upon public policies and implementing them in order to solve the problem of micro-pollutants in Rhine surface water.

Since our action situation focuses on policymaking and policy implementation, we consider insights from policy studies when analyzing actor collaboration and its outcomes. Policy studies emphasize collaboration mainly in politics and political decision-making and implementation. In this context, collaboration is heavily impacted by the salience of an issue or the seriousness of a problem: political actors put an issue on the agenda, negotiate about potential solutions to the identified problem, and consequently start to interact and collaborate as soon as problem identification is initiated (Jann and Wegerich 2003). The immediate products of collaboration are policies, most often taking the form of legal or binding decisions. In sum, to specify the action situation as outlined in the SES framework, we rely on policy studies and define it as "politics." The outcome of the action situation is typical public policies.

To study collaboration among the public and private actors involved in policymaking regarding micro-pollutant management in the Rhine catchment, we rely on tools from descriptive social network statistics. We assess and compare the structure of actor collaboration networks in three case studies in the Rhine river basin. To study actor collaboration thoroughly, we examine the networks at three levels: at the macro level, we look at the networks' cohesion and fragmentation; at the meso level, we scrutinize factions of the network; and at the micro level we focus on the integration of single actors.

The three case studies we assess are the sub-catchment of the Rhine River in the greater region of Basel, the Ruhr basin, and the Moselle basin on Luxembourgian and German territories. In all three regions, the problem of micro-pollutants in surface water is severe. However, the outcomes of actor collaboration, namely the public policies that tackle the problem of micro-pollutants, differ. In the Basel and the Ruhr case studies, measures are already implemented, while, in the case of the Moselle region, actors are at the stage of policy formulation. In this Chapter, we argue that differences in the intensity and structure of the collaboration networks, and thus diverse degrees of cohesion and fragmentation, can provide insights into the different results of cooperation in these cases.

The chapter continues with an overview of the topic's literature and the strands that inform our study. We then introduce our case studies and the data gathering procedure and outline the data analysis method, providing an introduction into descriptive Social Network Analysis (SNA). The section on the analysis summarizes the case studies' results, which we then discuss and relate to the bigger picture in the chapter's last section.

Literature

A considerable amount of studies investigates interactions of actors and the quality of interactions in order to study the process of solution finding to complex natural problems (Ostrom et al. 1994; Baland and Platteau 1996; Ostrom 2000, p. 148; Agrawal 2001; Araral 2014, p. 14). The SES framework is one way to frame how actors interact in order to solve an environmental problem (see Fig. 8.1). The resource system (RS) and the resource units (RU) represent the ecological system; the governance system (GS) and the actors (A) stand for the social system, with the

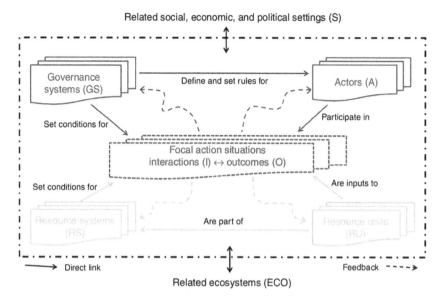

Related social, economic, and political settings (S)

Governance systems (GS)

Define and set rules for

Actors (A)

Set conditions for

Participate in

Focal action situations interactions (I) ↔ outcomes (O)

Set conditions for

Are inputs to

Resource systems (RS)

Are part of

Resource units (RU)

→ Direct link   Feedback - - - →

Related ecosystems (ECO)

Fig. 8.1 The social-ecological system framework (SESF). (Source: McGinnis and Ostrom 2014)

GS defining and setting the rules that A behave upon (Ostrom 2009; Basurto et al. 2013, p. 1367). The four elements all influence the action situation in which actors interact: The RS summarizes the ecological system under study. The GS stands for the policy area in which the environmental problem is addressed. The GS furthermore includes the institutions that set the legal rules, that structure the political system and guide its actions (Knill and Tosun 2012, p. 4, 41; McGinnis and Ostrom 2014), and the organizations that produce these rules. The RU includes the resource units that give the situational input to the action situation and substantiate the environmental problem at stake.

Collaborative governance emphasizes the interaction among diverse actors affected or concerned by a natural or common pool resource problem (Ansell and Gash 2008; Lubell et al. 2010). Those actors engage in collaboration—and should ideally benefit from it—to find a joint solution to the identified problem (Berkes and Folke 2002). Collaboration seems like a specific and unique type of interaction when it comes to solve a problem, as it needs the engagement of two or more actors in its creation and maintenance. This is different in the case of, for instance, information or resource exchange: information can be sent from a source or sender to a receiver. Only one party needs to engage in the creation of the relational tie. In the case of collaboration, though, both parties mutually contribute to the creation of this interaction to jointly fulfill a task or to solve a problem. This makes collaboration a strong tie (Ingold 2017) and might contribute to the creation of a network with stable interactions such as defined by Kenis and Schneider (1987). Actor collaboration can thus comprise creators and "victims" of an environmental problem who engage in a relation with each other. Furthermore, actors who create an environmental problem can also be affected by it themselves—for instance, an agricultural actor who applies pesticides that enter the same water cycle from which the actor obtains drinking water—and are thus interested in solving it through joint action.

In the action situations of our case studies, a broad array of actors come together to collaborate in order to produce beneficial outcomes that help solve the identified problem (McGinnis 2011; Fig. 8.1). On one side, we identify the actors affected or dependent upon the quality of the resource. This characteristic is informed by the resource system which we define as the river basin at stake, and by the resource unit, which we define as surface water contaminated with micro-pollutants. On the other side, there are actors responsible for the management and regulation of the resource surface water. These are informed by the governance system that is the sum of the rules and regulations that ought to manage the quantity and quality of surface water (Ostrom 2000; Schlager 2004).

We apply the logic of policy studies, which state that problems drive the agenda-setting and, finally, the production of public policies, to our conceptualization of the SES framework and state that the outcomes of the action situations we study are public policies, that is, official guidelines to solve the identified problem through binding targets and related policy instruments. We thus claim that these specific action situations of collaboration networks are policy networks that generate and implement public policies for water quality management. Our core assumption—which is confirmed by recent environmental policy studies (see Koontz and Thomas 2006; Ingold and Fischer 2014; Metz 2015; Newig et al.

2018; Yla-Anttila et al. 2018)—is that different types of collaboration networks lead to different types of outcomes in terms of public policies.

Case Description and Data Gathering

We focus on the environmental problem of micro-pollutants in surface water. Micro-pollutants are chemical substances that appear in concentration in the μg/l and ng/1 range in water bodies (Kümmerer 2009, p. 2354f.). Since their concentration is so low, they could not be detected until recently (Meyer et al. 2011, p. 128) and thus pose a fairly new environmental problem. Their sources are human-made and diverse: they stem from pesticide use in agriculture, herbicide applications on roofs, streets, and facades, use of pharmaceuticals and detergents, and industrial production processes. Micro-pollutants thus have point and diffuse entry paths which challenge their regulation. Public policies would need to rely on a wide portfolio of instruments in order to address all the different types of substances and to tackle their effects at their different entry points into the water system. The fact that micro-pollutants stay in water bodies—if they are persistent or are not filtered—, travel long distances and appear at different administrative levels—the local, the regional, the national, and even international—adds to the complexity of this environmental problem: different jurisdictions have to face it even though the problem's sources might lie outside their field of activity.

Studies on micro-pollutants raise the concern regarding their negative effects on ecosystems (Kümmerer 2009, p. 2360; Pal et al. 2010, p. 6063; Touraud et al. 2011, p. 437), their potential in creating joint toxicity when mixing (Kümmerer 2009, p. 2359; Touraud et al. 2011, p. 439), and the physiological changes they may induce in animals (Touraud et al. 2011, p. 437).

Micro-pollutants in water bodies thus fulfill the four criteria of a policy problem (Metz and Ingold 2014): they have different causes; they are omnipresent, with seasonal peaks regarding application periods in agriculture and run-offs in times of heavy rain; they show effects on flora and fauna; and they appear on different scales (cf. Herzog 2018, p. 43f.).

Three Case Study Areas and Research Design

We chose three case study areas along the Rhine. The criteria for their choice were (a) micro-pollutants had to have been detected in the river basin's water; (b) some body of regulations or management plans regarding the handling of the environmental problem had to be in place; (c) the river surface water had to be in use (e.g. for drinking water or fishing purposes); and (d) the regions had to be spatially disconnected to avoid a dependency of one case with another (Herzog 2018, p. 58, 67). The chosen case studies are the following:

- the River Rhine basin in the urban area of the canton Basel-City, Switzerland, which belongs to the hydrological sub-catchment High Rhine;
- the catchment area of the River Ruhr in the German federal state North Rhine-Westphalia (NRW); and
- the Moselle basin on the territory of Luxembourg and the two German federal states Rhineland-Palatinate and Saarland.

Regarding the environmental problem that provided the impetus for policymaking and collaboration in water quality management, actors in the three case study regions perceived micro-pollutants as equally severe; however, they put the issue on the political agenda at different points in time and with different aims.

In the Moselle case study, the chemical substances in the river's surface water stem from pharmaceuticals, agriculture, and winegrowing. Actors in this case study region started to collect information on the extent and the intensity of micro-pollutants in the river's surface water prior to any decision-making on source-directed or end-of-pipe measures. An incident of high concentrations of the herbicide metazachlor in the Moselle's tributary Sauer in 2014 (Luxemburger Wort, 29.09.2014) brought the attention of administrative actors and service providers in Luxembourg to the issue of micro-pollutants in river surface water. Consequently, water quality studies and pilot studies of filtering techniques at hospitals and selected wastewater treatment plants were done in Luxembourg and

Rhineland Palatinate to assess the pollution's intensity and extent and the performance of end-of-pipe measures. At the time of the data collection process, no measures to tackle the problem had been conclusively established in Luxembourg, while actors in the Moselle river basin on the German side had developed and implemented a set of source-directed measures, like regularly monitoring water samples to trace micro-pollutants' concentration and to identify dischargers. However, the actors in the German federal states Rhineland Palatinate and Saarland were still waiting for the EU guidelines on micro-pollutants before taking further measures (Herzog 2018, p. 80ff).

In the Ruhr case study, micro-pollutants are caused by settlements, waste water treatment plants, and agriculture. Public as well as private actors became active on the topic in 2006, when researchers detected high concentrations of perfluorooctanesulfonic acid (PFOS) in the Ruhr, which posed a serious threat to the region's drinking water source. In the following, federal state authorities developed a strategy plan for rivers and water bodies in North Rhine-Westphalia (NRW) with the aim to improve water body and drinking water quality. The program outlines a water monitoring system and water treatment practices, such as the upgrading of waste water treatment plants. Overall, actors in the Ruhr case study region follow a so-called multibarriers approach for micro-pollutant management, which implies that measures target the reduction of micro-pollutants at their source, the treatment of wastewater, and the processing of drinking water (Herzog 2018, p. 78).

In the Rhine catchment area at the city of Basel, micro-pollutants in the river's surface water stem from urban drainage, from the pharmaceutical and chemical industry in the city, and from the entries of micro-pollutants into the Rhine's tributaries on Swiss territory. Their load and concentration are constantly monitored at the international Rhine observation station at Weil am Rhein, a few kilometers downstream from Basel. An incident of chemical residues, at a waste disposal site close to Basel, that were leaking into the region's groundwater raised public attention on the chemical substances. In the aftermath, the city's two drinking water providers already upgraded their water processing techniques to filter remaining micro-pollutants. The city's waste water treatment plants (WWTP) will be upgraded before 2023 to eliminate micro-pollutants sufficiently. Apart from those end-of-pipe measures, regular water quality monitoring is installed, bans on certain substances are implemented, and compensation for preventive measures in agriculture is provided (Herzog 2018, p. 70ff, 87).

The seriousness of the problem and its salience is comparable across the cases: in the Ruhr and the Basel case, drinking water is provided by river bank filtration of river surface water; in the case of micro-pollutant management in the Moselle basin, the river is not used for drinking water, but does provide ecosystem services such as fishing. Actors in all three case study regions are aware of the water quality issues.

While the intensity of the problem is comparable across the cases, the outcome of politics in the action situation differs. In the Basel case study, policies are diverse and implemented; in the Ruhr case study, the multi-barriers approach is being implemented. In both cases, most of the policy instruments are of the "end-of-pipe" type and conceptualized as the technical solution of filtering micro-pollutants out of surface water. In the Swiss case, these policy instruments are guided by the national water protection act; in the Ruhr case, policies are realized through the federal water act and outlined in a strategy plan. In the Moselle case, actors start to gather information on the problem's extent and conduct feasibility studies of end-of-pipe measures. They have not yet produced a set of policy instruments that would tackle micro-pollutants at their source, or after they entered the water cycle.

The case studies represent thus different stages of instrument development and application in the context of water quality management in the Rhine basin. They further differ regarding the style of political decision-making in Switzerland, Germany, and Luxembourg, respectively—a characteristic of the case's Governance System (GS).

The three case study regions all lie in the Rhine basin, their sub-catchments constituting each case's resource system. The seriousness of the problem is similarly intense in all cases and the way in which the topic got on the political agenda is similar across the cases. The broader socioeconomic and institutional settings each water quality management is part of are similar and reflected in the governance system: the three case study regions are part of the territory of the International Commission for the Protection of the Rhine (ICPR), and water quality management in all three regions is subject to the EU Water Framework Directive (WFD).[1]

Table 8.1 summarizes the case studies' characteristics, their action situation, and the policy outcomes.

Policy Network and Actor Identification Therein

This study examines whether the intensity and structure of collaboration among actors engaged in an environmental problem can provide insights on the difference in policy outcomes. We claim this actor collaboration to happen within a policy network. A policy network is a set of public and private actors interested in finding a public solution to an identified problem, and therefore engaging in a set of stable interactions among each other (see Kenis and Schneider 1987). The type of interaction we are interested in is collaboration, as we consider collaboration as the stable and mutual interaction of actors when negotiating, introducing, and implementing public policies to address environmental problems such as the issue of micro-pollutants. To identify actors, we relied on the SES framework (Ostrom 2005), which provides actor characteristics that guide the identification process, and on the reputational approach (Knoke 1998). We further followed Mayntz and Scharpf's (1995) premise that a broad array of actors (in contrast to authorities or the iron triangle, see also Sabatier and Jenkins-Smith 1993) is involved in today's policymaking. The documents used to identify the actors were official documents, reports, and studies about water quality policymaking in the regions, as well as homepages of water-related associations, resource users, and political authorities.

We identified actors that depend upon or use the resource surface water, that is, service providers, consumers, and polluters; authorities from different levels; scientific experts and NGOs; and consumer and water associations. The actors' list derived from this structured search was validated by experts. The chapter's first author interviewed 22 experts

---

[1] Switzerland is not part of the European Union, and is thus not obliged to follow the EU WFD. However, the country calibrates its actions to the WFD, in part because the ICPR adapted its monitoring program according to the WFD, and Switzerland is an ICPR member state.

Table 8.1 Overview of the case studies' characteristics, the actor collaborations' state, and the policy outcomes

Case study	Resource system (RS)	Environmental problem	Governance system (GS)	The trigger for the environmental problem to be set on the agenda	Outcome: public policies
Basel	Transboundary Rhine catchment area at the city of Basel	Micro-pollutants in Rhine surface water	Swiss political system; part of the territory of the international commission for the protection of the Rhine (ICPR)	Based on water quality monitoring and incident of chemical residues in groundwater	Monitoring; research; implementation of source-directed and end-of-pipe measures
Ruhr	Sub-catchment area of the Rhine tributary Ruhr	Micro-pollutants in Ruhr surface water	German political system; water policy of North Rhine-Westphalia (NRW); part of the ICPR's territory;	Detection of PFOS in high concentration	Monitoring and multi-barriers approach: source-directed measures; end-of-pipe measures (4th treatment stage for WWTP; processing of drinking water)
Moselle	Transboundary sub-catchment area of the Rhine tributary Moselle on German and Luxembourgian territory	Micro-pollutants in Moselle surface water	Luxembourgian political system and water policy of Rhineland Palatinate and Saarland; part of the ICPR's territory;	Agricultural accident causing high concentrations of the herbicide metazachlor in tributary of the Moselle	Water monitoring and feasibility studies on sewage treatment.

prior to the data gathering process (cf. Herzog 2018, p. 275f.). The experts indicated whether all actors important to the respective case study region were included in the list—and if not, which actor was missing—the procedure reflecting the reputational approach (Abu-Laban 1965, p. 35f.; Scott 2000, p. 56).

The actor identification process resulted in 44 collective actors in the Moselle case, 39 in the Ruhr case, and 51 in the Basel case (cf. Herzog 2018, p. 101ff). They are private and public entities from science, the industry, the civil society, the water sector, and the state. The proportion of the different actor types is quite similar across the three case studies.[2] In the cross-national case study, the German actors (38.6%) are outnumbered by the Luxembourgian actors (59.1%).

Collecting Network Data Through Surveys

Data was collected through a survey that was sent out to the actors via e-mail and post in the spring and fall of 2016. In each organization, the questionnaire was addressed to the individual in a key position regarding the issue of micro-pollutants. Such key positions comprised heads of department and working groups, professors, and CEOs (cf. Herzog 2018, p. 95).

The network data on actors' collaboration were collected by the following question:

> *With which actors has your organization been closely collaborating within the management process of micro-pollutants throughout the last years?*[3]

---

[2] There is an exception regarding NGOs and scientific actors: In the Basel case study, five actors are NGOs (accounting for 9.8% of the case's actor sample), while, in the Ruhr case study, only one actor is an NGO (constituting 2.6% of the actor sample). Scientific actors diverge even more, accounting for 17.6% (nine actors) in the Basel case study, to 23.1% (nine actors) in the Ruhr case study, and 9.1% (four actors) in the Moselle case study.

[3] To assure actors understand the same by collaboration, a brief description of which actions can potentially fall under the term collaboration was stated below the question: discussing new findings on the issue, working out possible courses of action regarding the management of micro-pollutants, exchanging viewpoints on the topic, and accomplishing joint projects regarding micro-pollutants.

In a roster of the list of actors involved in the survey, actors could mark with a cross those actors that they collaborate with. Answers were then converted into a matrix (cf. Borgatti et al. 2013, p. 63f.): a 1 in a cell indicates the two actors belonging to the cell collaborate while a 0 indicates no collaboration. These network matrices provide the database for the subsequent network analysis. Note that even if we conceptually consider collaboration as a mutual engagement in an interaction, we did not symmetrize the data: if actor A mentioned collaborating with actor B, but this tie was not confirmed by B, we did not, ex-post, reciprocate this tie. The reason for that decision on our part is that perceived collaboration can mean different things to two actors or organizations.

The response rates reached 66.6% in the Ruhr case study (i.e. 26 actors), 70.5% in the Moselle case study (31 actors), and 72.5% in the Basel case study (37 actors). A look at the actors considered important by their peers (separate survey question) reveals that all the actors considered important by those who answered the survey in the Basel case study did answer the questionnaire, and that their share is 71.4% in the Ruhr and Moselle case studies.[4]

The response rates are thus satisfactory, with the vast majority of each case study's key actors responding to the survey, ensuring that our study includes the correct actors for each case. Each actor group is represented in the case studies, with the Basel case study having the most regular distribution of actor types (Herzog 2018, p. 101ff.).

Methods of Descriptive Social Network Analysis (SNA)

To identify potential differences in the collaboration networks across the three cases, we rely on descriptive Social Network Analysis (SNA). We focus on two prominent network concepts, cohesion and fragmentation, to assess the quality of collaboration. Cohesion can be defined as the

---

[4] The benchmark, by how many actors an actor has (?? Do you mean how many connections an actor has?) to be judged as important to regard him/her important for this analysis, was set at 40% for each case study's actors.

degree of connectivity in a graph (Valente 1996; Reffay and Chanier 2003). One basic, constituent function of a network is to link nodes so that the network does not fall apart. Cohesion represents the connectedness of nodes that keeps the network together.

A network's connectedness is "the proportion of pairs of nodes that are located in the same component" (Borgatti et al. 2013, p. 154). The counterpart of a network's connectedness is its fragmentation (see Angst et al. 2018): it reflects the absence of ties in a network and the subgroups of actors or nodes only loosely connected to the rest. Fragmentation and connectedness are complementary[5] and reflect a network's cohesion from two different angles.

When assessing actor collaboration with Social Network Analysis, the cohesion of the collaboration network is treated as an indication of the collaboration's intensity. A cohesive network consists of a high number of collaboration ties among the actors present in the network. There are few subgraphs, and the tendency is for one large component to dominate the network. If there are subgraphs in a cohesive network, few central actors are able to link those together. A fragmented network looks differently: it is characterized by many loosely connected subgraphs, components, or factions. And it is always different actors that link these subgraphs together and to the rest of the network. In the following, we present different measures of cohesion and fragmentation at the macro (overall network), meso (subgraphs), and micro (node) levels of the case studies' collaboration networks.

Edges and Vertices That Make a Graph

A social network is made up of vertices (also called nodes) and the links between these vertices, the edges (also called ties or relations). Edges represent thus the vertices' relations. The set of vertices and the set of edges constituting the network are called a graph. A graph is the network's mathematical representation (Borgatti et al. 2013, p. 11f.).

---

[5] Connectedness equals 1 minus fragmentation (cf. Borgatti et al. 2013, p. 154).

A dyad is the relation between two vertices: the "pair of actors and the (possible) tie(s) between them" (Wasserman and Faust 1994, p. 18). Two actors that are connected through a tie are said to be *adjacent* to each other (Wasserman and Faust 1994, p. 95; Scott 2000, p. 67; Borgatti et al. 2013, p. 12). The type of actors' relations can be manifold: they can be communication, resource or information exchange, a biological or friendly relationship, a shared point of view, the mutual perception of each other, or a behavioral interaction (Wasserman and Faust 1994, p. 18; Schneider 2014, p. 274). In our case, we focus on collaboration exclusively.

Vertices have also characteristics that distinguish them from each other. Such characteristics can be a person's sex, age, or profession, an organization's ideology, or a nation's CO_2 emissions. These characteristics are the nodes' attributes, which are treated like variables with different values. These are saved as attribute data (Scott 2000, p. 2). The only attribute data we are interested in in our case studies are actor types. The actor samples comprise eight types of actors: NGOs, consumer groups, science, water associations, service providers, polluters, national state actors, and regional state actors.

Cohesion and Fragmentation: Macro-, Meso-, and Micro-level Statistics

We operationalize and assess the intensity and structure of the case studies' collaboration networks with different network measures. They are outlined below, and classified following their unit of analysis (network, subgraph, actor).

Macro Level

At the network's macro level, we focus on density and reciprocity to indicate the degree of cohesion. Density is one of the most prominent measures at the network level. It is "the probability that a tie exists between any pair of randomly chosen nodes" (Borgatti et al. 2013, p. 150). It is "the share of ties existent within a network compared to the amount of ties theoretically possible within this network" (Herzog 2018, p. 110). The density's value of a directed graph reaches from 0 to 1: the value of 0 means that there are no edges; at the value of 1 all possible ties are present (Wasserman and Faust 1994, p. 129).

Reciprocity reflects the number of ties that are being reciprocated—that is, when actor A sends a tie to actor B and actor B confirms the existence of this tie. As mentioned above, in this research, we treat collaboration as a perceived tie: it is thus possible that two survey partners in our case do not evaluate the interaction among each other in the same way, which can result in an unreciprocated tie between the two (when one actor indicates collaborating with the other, but the second actor does not confirm this). A network with a high reciprocity value reflects a cohesive network in that its reciprocated ties represent a specifically strong connection between two nodes.

Meso Level

At the meso level, we look at subgraphs of the network. Generally speaking, the more subgraphs we identify, the higher the fragmentation and thus the lower the cohesion in that collaboration network. We measure two types of subgraphs here: components and factions.

A component represents a subgraph whose vertices are reachable for every other vertice in this subgraph (Scott 2000, p. 101). No path[6] towards any vertice outside this subgraph exists (cf. Wasserman and Faust 1994, p. 109; Scott 2000, p. 101). For directed networks, one distinguishes *weak* and *strong* components: strong components recognize the direction of ties, weak components do not (Scott 2000, p. 103f; Borgatti et al. 2013, p. 16). A network that has several components is less dense and more fragmented, and a network made up of one large component (and only few isolates[7] or smaller components) is dense and cohesive (Scott 2000, p. 104). A graph that consists of one component is

---

[6] A path is a sequence of connections in which both—edges and vertices—are only included once (Scott 2000, p. 68).

[7] Isolates are nodes that have no connections (Borgatti et al. 2013, p. 14).

connected; a graph with more than one component is disconnected (Wasserman and Faust 1994, p. 109).

To check for subgroups of actors within a network, we do a faction analysis. Factions are "cohesive groups of nodes whose number is predetermined" (Borgatti et al. 2013, p. 191; Herzog 2018, p. 109). Actors within one faction have to be adjacent or linked through a path (Wasserman and Faust 1994, p. 290). Faction analysis is explorative: one determines the number of factions that vertices are grouped into *before* the analysis is executed. Each node is assigned to one faction only.[8] One has to repeat the analysis several times to assure that nodes are constantly assigned to the same faction and guarantee the faction partition's robustness (Borgatti et al. 2013, p. 192).

Micro Level

The micro level focuses on single vertices, that is, the actors in the network. The position of the vertices and the connections they possess inform us about how cohesion plays out on the micro level. The more actors of the network dispose high centrality measures, as, for instance, degree centrality, the more connected and cohesive the network is. Moreover, further fragmentation of a network can be prevented by those actors that have a high betweenness centrality (see below) and that connect a network's different subgraphs. Centrality measures conceptualize the vertices' positions within a network. The three most prominent centrality measures are degree, closeness, and betweenness centrality. For our study, we focus on betweenness centrality only.[9]

Betweenness centrality focuses at a vertice's capacity to link other vertices of the network, measuring the proportion of the shortest paths of all pairs of vertices that pass through the focal vertex, and then summing

---

[8] In the case that a node entertains paths or is adjacent to one or more nodes in several factions, the algorithm still forces it into one faction only (cf. Herzog 2018, p. 109).

[9] For more details on degree centrality, see Wasserman and Faust (1994, p. 101, 126, 178f), Scott (2000, p. 83) and Borgatti et al. (2013, p. 165f.).

these proportions up to obtain a single value for the focal vertice (Wasserman and Faust 1994, p. 188ff; Borgatti et al. 2013, p. 174f; cf. Herzog 2018, p. 110).[10]

Analysis

Before analyzing the intensity and structure of the collaboration networks in the Basel, the Ruhr, and the Moselle case study along the measures outlined above, we provide the general statistics for the networks under study.

The actor network in the Basel case study is the largest with 37 actors (see Table 8.2). The actor networks of the Moselle and Ruhr case studies have 31 and 26 actors, respectively. The number of edges varies according to the number of vertices in the three cases: the network of the Basel case study has 276 ties, while the one of the Moselle case study has 216, and that of the Ruhr case study comparatively less ties, that is, 160.

Despite their difference in ties, the networks differ only slightly with regard to the average degree of ties their vertices possess. In the Basel case, an actor holds 7.5 ties on average; in the Moselle case 7, and in the Ruhr case 6.2 ties. The standard deviation is approximately the same in all networks, indicating "that the error rate of estimating a tie between two nodes in the networks is at 40%" (cf. Herzog 2018, p. 121).

Table 8.2   Basic network statistics of the case studies' collaboration networks[a]

Statistics	Basel	Ruhr	Moselle
Nodes n	37	26	31
Ties	276	160	216
Average degree	7.459	6.154	6.968
Standard deviation	0.405	0.431	0.422

[a]Network statistics were calculated in UCINET

---

[10] For an extensive examination of centrality measures, see Chap. 5 in Wasserman and Faust (1994) and Chap. 10 in Borgatti et al. (2013).

The Macro Level: Density, Reciprocity, and Connectedness Statistics

At the macro level, we look at the networks' density, reciprocity, and their connectedness. To compare networks of different sizes, one can refer to their density since density "(…) adjusts for the number of nodes in the network, making density figures comparable across groups of different sizes" (Borgatti et al. 2013, p. 151). It sounds reasonable that in a smaller network it is easier to reach out to a lot of actors than in a large network with many actors. Densities are thus "(…) almost always lower in large networks than in small networks" (ibid., p. 151). This is also the case for this study: we observe the lowest density of 20.7% in the largest collaboration network (Basel case study) and the highest density of 24.6% in the smallest network (Ruhr case study), while the Moselle case study's collaboration network has a density of 23.2% (see Table 8.3). Overall, the networks are similarly dense.

Reciprocity of ties is almost even in the three networks, ranging from 28.9% of reciprocated ties in the Basel case study to 29.4% in the Ruhr and 29.6% in the Moselle case study. With respect to their density and reciprocated ties, the three observed collaboration networks are akin to each other. Regarding their connectedness measure, the networks are also similar: in the Ruhr case's collaboration network: 85% of actor pairs are connected. This share is even higher in the other two networks, amounting to 92% in the Basel case study and 94% in the Moselle case study. Fragmentation—defined as 1 minus the value of connectedness—is accordingly low in the networks: 6.3% and 8% of actor couples in the Moselle and the Basel case studies' collaboration networks do not reach out to each other. In the Ruhr case study, this share is higher: 15% of potential dyads do not share a tie.

The networks' high cohesion, defined by the networks' densities, their reciprocated ties, and their connectedness measure, also shows in their

Table 8.3 Network statistics of the collaboration networks

	Basel	Ruhr	Moselle
Mutual ties	80	47	64
Reciprocity	0.2899	0.2938	0.2963
Density	0.207	0.246	0.232
Connectedness	0.92	0.851	0.937
Fragmentation	0.08	0.149	0.063
Normed mean of reach centrality	0.54	0.54	0.55

reach centrality. Reach centrality refers to the number of actors an actor reaches out to in 1, 2, 3, or more steps—a step being an edge between two vertices (Hanneman and Riddle 2005, Chap. 10). This value is even and high in all three networks: in the Basel and Ruhr case studies' networks, an actor connects on average to 54% of the network's actors with only one step; in the Moselle case study's network the value is 1% higher. On average, actors in all three networks are linked to more than half of the respective network's actors by just one tie; a fact that underlines the networks' strong connectedness and cohesion. The study of the networks' subgraphs shows a more detailed picture of cohesion and fragmentation in the three networks and the differences among them.

The Meso Level: Components and Factions

A component is a network's subgraph in which all actor pairs are linked with each other through a path (Herzog 2018, p. 129). The networks of the Basel and Ruhr case studies have four, the network of the Moselle case study has three strong components (see Table 8.4). All networks possess one exceptionally large component consisting of between 89% and 94% of the networks' actors, while the remaining components are made up of only one actor each.[11] These large components further express the networks' strong cohesion, because "(…) the bigger the main component (in terms of nodes), the greater the global cohesion of the network" (cf. Borgatti et al. 2013, p. 153).

Table 8.4 Collaboration networks' strong components

No	Basel		Ruhr		Moselle	
	Size	Share	Size	Share	Size	Share
1	34	0.919	23	0.885	29	0.935
2	1	0.027	1	0.038	1	0.032
3	1	0.027	1	0.038	1	0.032
4	1	0.027	1	0.038		

[11] The Ruhr case study's network also has two weak components, with one representing an actor that is completely disconnected from the rest of the network. These components are not shown in Table 8.4.

The existence of components in networks can be represented with the component ratio[12]: if the network possesses only one component, the component ratio equals the value of 0; if every node in the network is an isolate, the value of the ratio is 1. Subtracting a network's component ratio from 1 produces the network's cohesion measure (Borgatti et al. 2013, p. 153). Since the three collaboration networks each have one large component and only few small ones, their component ratios are low[13] while their cohesion measures are high: 0.933 for the Moselle case's network; 0.917 for the Basel case's network; and 0.88 for the Ruhr case's network. The component analysis thus further informs us about the collaboration networks' cohesion.

By conducting a faction analysis, one breaks up the big components in each network, and identifies a network's subgroups. The faction analysis further specifies whether these subgroups build along specific actor characteristics. The faction analysis[14] of the Basel case study consists of four factions (see Table 8.5), three of which are heavily dense. In the first faction of 12 actors and a density of 53%, service providers and scientific actors come together. The second faction of ten actors and a density of 61% is the one where regional political actors and polluters meet. In the third faction, nine actors mainly from the civil society and national water associations group, their tie density being 51%. In the fourth faction, French, Swiss, and German actors from the political sector, the civil society, the industry, and the service sector have only few connections (density: 3%).

The factions in the Ruhr case study's collaboration network divide along the service sector on the one side and scientific and political actors on the other (Table 8.6). The first faction is made up of the regional scientific and political actors, as well as a regional water association, a drinking water provider, and an environmental NGO; the eleven actors

---

[12] The component ratio is the number of components, c, minus 1 divided by the number of nodes in the network, n, minus 1: $c - 1/n - 1$ (Borgatti et al. 2013, p. 153).

[13] The component ratio values for the three networks are 0.067 for the Moselle case study, 0.083 for the Basel case study, and 0.12 for the Ruhr case study.

[14] To ensure the validity of the faction analysis' results, the algorithm was run several times. This procedure assured that actors were always assigned to the same faction. The factions' number, that needs to be determined *prior* to the analysis, was decided for each case study based on a series of different numbers of faction partitions that were run beforehand. Based on their validity in actor assignment, their value in meaningfulness, and their final proportion correctness, the number of factions for each case study was chosen (cf. Herzog 2018, p. 134f.).

Table 8.5 Faction analysis of the Basel case study—final proportion correctness: 79.4% (actor types are highlighted in same colors)

	Faction 1		Faction 2		Faction 3		Faction 4	
	density: 53%		density: 61%		density: 51%		density: 3%	
	Actor	Type	Actor	Type	Actor	Type	Actor	Type
1	AWBR	Water association	AIBBL	Regional polit. actor	ALSACE NAT	NGO	ADM Loerr	Regional polit. actor
2	IAWR	Water association	AUEBL	Regional polit. actor	AQUA VIV	NGO	CITY Weil	Regional polit. actor
3	SVGW	Water association	AUEBS	Regional polit. actor	PRONA	NGO	KI.SS V. SGV	Regional polit. actor
4	EAWAG	Science	WWTP Rhein	Polluter	WWF	NGO	APRO NA	NGO
5	TZWK	Science	WWTB ChemBasel	Polluter	VSA	Water association	HKBB	Polluter
6	AQUA EXP	Science	NOVAR TIS	Polluter	ICPR	Water association	SBrV	Service provider
7	FSVO	National polit. actor	ROCHE	Polluter	CERCL	Science		
8	LABBL	Regional polit. actor	WWTP Basel	Polluter	SFA	Consumer organization		
9	LABBS	Regional polit. actor	WWTP Birs	Polluter	FOEN_W	National polit. actor		
10	WWR	Service provider	KI	Service provider				
11	IWB	Service provider						
12	WWB	Service provider						

Table 8.6 Faction analysis of the Ruhr case study—final proportion correctness: 73.8% (actor types are highlighted in same colors)

	Faction 1		Faction 2		Faction 3	
	density: 59%		*density: 52%*		*density: 12%*	
	Actor	Type	Actor	Type	Actor	Type
1	UNI.Boch	Science	ARW	Water association	BMG	National polit. actor
2	UNI.Duis	Science	AWWR	Water association	UBA	National polit. actor
3	CompCent. NRW	Science	VKU. NRW	Service provider	Fish. NRW	Consumer organization
4	RWTH Aach	Science	WWW	Service provider	Fish. RUHR	Consumer organization
5	Eawag	Science	DEW21	Service provider	Paper. NRW	Polluter
6	IWW	Science	Gelsen. plc	Service provider	RLV	Polluter
7	MKULNV	Regional polit. actor	InstHyg. Gelsen	Science	WLV	Polluter
8	DA.Duess	Regional polit. actor	CoA. NRW	Polluter		
9	BUND. NRW	NGO				
10	RWW	Service provider				
11	RV	Polluter				

connect with a density of 59%. In the second faction, eight actors from mainly the water service provision sector entertain ties at a density of 52%. The third faction consists of agricultural and industrial polluters, fishing associations, and two national state actors who are more loosely connected (density: 12%).

Table 8.7 Faction analysis of the Moselle case study—final proportion correctness: 81.3% (actor types are highlighted in same colors)

	Faction 1		Faction 2		Faction 3		Faction 4	
	density: 67%		*density: 60%*		*density: 57%*		*density: 5%*	
	Actor	Type	Actor	Type	Actor	Type	Actor	Type
1	MinAgri. LUX	National polit. actor	SGD	Regional polit. actor	EVS	Polluter	FLPS	Consumer organization
2	OffNat. LUX	National polit. actor	OffNat. RLP	Regional polit. actor	SIDEST	Polluter	ULC	Consumer organization
3	MinDev. LUX	National polit. actor	MUEEF. RLP	Regional polit. actor	MUV. SAAR	Regional polit. actor	SIVEC	Polluter
4	City. LUX	Regional polit. actor	MWVLW. RLP	Regional polit. actor	OffNat. SAAR	Regional polit. actor	Hospitals .LUX	Polluter
5	CoA. LUX	Polluter	LDEW	Water association	Fish. SAAR	Consumer organization	StGB.RL P	Regional polit. actor
6	SIDEN	Polluter	CoA.RLP	Polluter	TUKais	Science		
7	SIDERO	Polluter	SWT	Service provider	UNI. LUX	Science		
8	SEBES	Service provider						
9	SES	Service provider						
10	Natur Emw	NGO						
11	LIST	Science						
12	Aluseau	Water association						

In the transboundary collaboration network of the Moselle case study, actors group in factions along their nationalities (see Table 8.7). The first faction consists entirely of Luxembourgian actors from the political, the service provision, the polluting and scientific sectors, and the civil society.

The connections of these 12 actors have a density of 67%. The second faction is made up of actors from the German federal state Rhineland-Palatinate (RLP), coming from the polluting, the political, and the service provision sector (density of 60%, $n = 7$). The third faction also has seven actors who mainly come from the German federal state Saarland (two scientific actors are Luxembourgian and from RLP). This faction's density is also high (57%). The last faction of five actors comprises consumer organizations and polluters from Germany and Luxembourg, and has a low density of 5%.

The faction analysis sheds light on how differently actors group together in the three case studies. The networks' densities, tie reciprocity, and components showed a homogenous picture of cohesive collaboration networks. The composition of the factions shows differences between the three networks. In the Basel case study, the actors' background appears to structure the factions: service providers together with scientific actors form a subgroup, while regional polluters form a faction with political actors; and civil society collaborates strongly with their peers. In the Ruhr case study, actor type structures the factions as well. Here, however, the scientific actors seek collaboration not with service providers, but with the political actors, while actors from the service provision sector form a collaborating subgroup on their own. Collaboration in factions in the Moselle case study is divided along nationalities, with cross-border collaboration between actors from Saarland and Luxembourg.

The Micro Level: Actors in Between: Building Bridges Across Factions

The centrality measure we focus on is betweenness centrality, since it represents the number of shortest paths between pairs of actors that pass a node[15]: An actor with a high betweenness centrality is thus an actor through whom many actor pairs are connected. Actors with a high betweenness centrality in the three studied collaboration networks are those who can function as a bridge, and/or a gateway keeper (Jasny and Lubell 2015, p. 38) between actors who are not directly linked.

---

[15] Centrality measures were calculated in UCINET, using Freeman Betweenness Centrality.

In all three case studies, the actor with the highest betweenness centrality is the national or federal ministry for the environment (FOEN_W, MKULNV, MUEEF.RLP; see Table 8.8). In the Basel case study, the following actors with high betweenness centralities are two regional political actors who guide water quality policy in Basel and its surroundings and a Swiss and a German scientific actor. Interestingly, the scientific actors (Eawag and TZWK) are both in faction 1, the regional political actors (AUE BL and AUE BS) are both in faction 2, and FOEN_W is part of faction 3 (see Table 8.5). The bridging actors are thus distributed across the collaboration network's three main factions[16] and provide the connections between actor pairs, and most likely also across the factions.

In the Ruhr case study, a polluter (RV), a regional water association (AWWR), the German ministry for the environment (BMG) and a scientific actor (IWW) are the actors following the MKULNV in high betweenness centrality. The federal ministry for the environment is in a faction together with RV and IWW (Faction 1, see Table 8.6) while AWWR is in faction 2 and the BMG in faction 3. In this case study as well, the bridging actors with the highest betweenness centrality sit in different factions, allowing for actor pairs to connect through them, thereby enhancing the network's cohesion.

In the Moselle case study, it's the Luxembourgian ministry for the environment (MinDev.LUX) and the one from Rhineland Palatinate (MUEEF.RLP) that have the highest betweenness centrality in the actor network. They are located in different factions, the Luxembourgian ministry being in faction 1, the entirely Luxembourgian faction; the environmental ministry of RLP being part of faction 2, the entirely German faction (see Table 8.7). The polluter Aluseau, which also has a high betweenness centrality value, is in the same faction as the Luxembourgian ministry of the environment; and so is the scientific actor LIST, which scores high in betweenness centrality as well. The universities in RLP and Luxembourg (TUKais and UNI.LUX) that possess a considerably high betweenness centrality are part of the third faction, thus enabling connections between German and Luxembourgian actors in the collaboration network.

---

[16] Faction 4 has a density of 3%, its members thus entertaining loose connections.

Table 8.8 Actors' normed betweenness centrality—the samples' 40% of actors with the highest betweenness centrality (actor types are highlighted in same colors)

Basel			Ruhr			Moselle		
Actor	**Sector**	**nBetw.**	**Actor**	**Sector**	**nBetw.**	**Actor**	**Sector**	**nBetw.**
FOEN_W	national pol. actor	27.673	MKULNV	regional pol. actor	29.323	MUEEF. RLP	regional pol. actor	16.954
AUEBL	regional pol. actor	10.222	RV	polluter	10.941	MinDev. LUX	national pol. actor	13.967
EAWAG	science	9.214	AWWR	water assoc.	7.030	Aluseau	polluter	11.431
AUEBS	regional pol. actor	8.759	BMG	national pol. actor	6.796	LIST	science	11.166
TZWK	science	6.719	IWW	science	4.355	TUKais	science	10.946
WWF	NGO	5.342	Gelsen.plc	service provider	2.992	UNI.LUX	science	6.672
SVGW	water assoc.	5.306	ARW	water assoc.	2.769	MUV.S AAR	regional pol. actor	6.091
LABBL	science	5.297	RWTHAach	science	2.433	MinAgri .LUX	national pol. actor	5.992
ICPR	water assoc.	3.496	DA.Duess	regional pol. actor	1.915	OffNat.L UX	national pol. actor	5.496
LABBS	science	2.945	UNI.Duis	science	1.184	SES	service provider	4.582
ESPECS	national pol. actor	2.918				SGD	regional pol. actor	4.519
WWB	service provider	2.750				SIDEST	polluter	3.370
WWTP Rhein	polluter	2.180						
WWTP ChemBasel	polluter	1.994						
WWTP Basel	polluter	1.994						

The look at the actors with a high betweenness centrality value in the collaboration networks showed that these are mainly political and scientific actors who are part of different factions, and may thus enable connections between actor pairs within, as well as across, factions.

Discussion and Conclusion

We chose three sub-catchments of the River Rhine to study actor collaboration around the environmental policy problem of micro-pollutants in surface water. We examined collaboration through Social Network Analysis to assess its intensity and structure across the three cases. In this conclusion, we discuss whether differences in actor collaboration may account for differences in the collaborations' outcomes, that is, the policies aiming to solve the environmental policy problem in question.

Both the environmental problem itself, and the pressure the problem puts on actors under study is similar in all three cases. However, the policy outcomes in the three case study regions are different: while in Basel, different technical measures and monitoring are already implemented, in the Ruhr case study, such policy instruments were introduced more recently. In the Moselle case study, no comparable policy instruments to fight micro-pollutants have been introduced yet.

Our results show that the collaboration networks in all three cases are similar: the number of observable ties (density), the way how actors engage in mutual relationships (reciprocity), the networks' components, reach centrality, and connectedness are very comparable across the cases.

Given all these indices, cohesion is high in all three networks. The degree of cohesion itself is comparably intense among the three regions. Since these measures do not differ across the cases, they do not provide explaining factors for potential differences in the outcomes of collaboration (Gerring 2001, p. 210; Bennet 2004, p. 31).

The major difference lies in the networks' factions. The faction analysis has shown that actors group differently in cohesive subgroups across the cases. In the case of the Moselle region, actors form factions along their territoriality. In the Ruhr case study, there is a divide between sectors, with service providers located on one side, and scientific and regional state actors on the other. This structural separation illuminates the line of conflict between these actors regarding the implementation of end-of-pipe measures. The Basel case shows a division that occurs roughly according to specialization: strong collaboration can be observed among actors from the civil society, among service providers with scientific actors, and among polluters with regional state actors.

Reflecting on actor collaboration's impact on policy outcomes, our results suggest that territorial and thus country-specific network fragmentation makes the production of coherent policy outcomes a difficult task. In Luxembourg, where cross-country collaboration is rather weak, we observe a "border effect" (Sohn et al. 2009). This fact might have hindered the introduction of policies to address the *transboundary* issue of micro-pollutants. Cross-sectoral fragmentation, in turn, does not seem to hamper the production of policy solutions. However, in the Ruhr case, where there is a slight sectoral fragmentation, we observe a slightly reduced bindingness of policy measures related to micro-pollutants. The Basel case's collaboration network is the most multi-level, cross-sectoral, and transboundary one of the three cases, which might have impacted the policies.

To some extent, the fact that the collaboration networks are similar can speak to the literature about problem pressure and salience. Micro-pollutants became a salient issue in all three case study regions, and therefore seemingly forced public and private actors to coordinate action to tackle it. The difference in policy outcome might well be a result of the maturity of the processes: in Switzerland, there was a push from research into politics in favor of the environmental issue. New techniques for upgrading treatment plants were developed hand in hand with new results and evidence on the impacts, the extent, and potential consequences of micro-pollutants. This certainly accelerated the process of making policy decisions. In the Ruhr basin, alarming evidence about water quality impacted policymaking regarding micro-pollutants. Whereas in the Moselle region, neither incentives from science nor tremendous incidents happened, and coordinated studies and action taking were missing. From this we can conclude that the push from science (Moraes et al. 2019), or the impact of focusing events (Birkland 2005) seem to have an impact on the policy outcomes as well.

Our results offer further interesting insights into policymaking in the Rhine catchment, and into water quality management in general. Interestingly, always the same type of actors—science and authorities—play the role of so-called brokers with high betweenness centrality. These actors are central because they figure as "leaders" within their faction, but have also the potential to link actors from other factions and of different organizational type (see Angst and Fischer, this book; Angst et al. 2018). Several reasons might account for the fact that science and authorities play this "connector" role. It could be that there is a general pattern of "neutral" scientist or public authorities playing the role of brokers in a policy process (Ingold and Varone 2012; Lu 2015). Through the information they provide and the role they play—that of an expert or a coordinator—they might be seen as a credible source of knowledge and power, which creates a reason for other actors to connect with them.

Another explanation for why these actors play prominently within the collaboration networks may lie in the nature of the subsystem: water quality issues, and the management of micro-pollutants in particular, are very technical and not particularly ideological topics. These issues asked for the expertise from science and the experience from authorities. Collaboration patterns and the structure of the policy network may be less influenced by the political styles of the countries, and even more so by the subsystem characteristics (Cairney et al. 2016).

Finally, the collaboration networks in the three catchments might look alike because of the transboundary learning within the Rhine catchment. The International Commission for the Protection of the Rhine (ICPR) is a well-established and organized forum, where public and private actors, experts, and consultancies exchange information and insights and streamline their actions regarding different water topics, that is, water quality issues such as micro-pollutants. The similar set-up of collaborative ties in a multi-level and cross-sectoral manner in all three case studies might be the result of joint venue participation and forum membership (Lubell et al. 2010; Lubell 2013; Fischer and Leifeld 2015; Herzog and Ingold 2019).

References

Abu-Laban, B. (1965). The Reputational Approach in the Study of Community Power: A Critical Evaluation. *The Pacific Sociological Review, 8*(1), 35–42.

Agrawal, A. (2001). Common Property Institutions and Sustainable Governance of Resources. *World Development, 29*(10), 1649–1672.

Angst, M., Widmer, A., Fischer, M., & Ingold, K. (2018). Connectors and Coordinators in Natural Resource Governance: Insights from Swiss Water Supply. *Ecology and Society, 23*(2), 1.

Ansell, C., & Gash, A. (2008). Collaborative Governance in Theory and Practice. *Journal of Public Administration Research and Theory, 18*(4), 543–571.

Araral, E. (2014). Ostrom, Hardin and the Commons: A Critical Appreciation and a Revisionist View. *Environmental Science and Policy, 36*, 11–23.

Baland, J., & Platteau, J. (1996). Halting Degradation of Natural Resources: Is There a Role for Rural Communities? Retrieved January 26, 2020, from http://www.fao.org/docrep/x5316e/x5316e00.htm.

Basurto, X., Gelcich, S., & Ostrom, E. (2013). The Social-ecological System Framework as a Knowledge Classificatory System for Benthic Small-scale Fisheries. *Global Environmental Change, 23*(6), 1366–1380.

Bennet, A. (2004). Case Study Methods: Design, Use, and Comparative Advantages. In D. F. Sprinz & Y. Wolinsky-Nahmias (Eds.), *Models, Numbers, and Cases: Methods for Studying International Relations* (pp. 19–55). Michigan, MI: University of Michigan Press.

Berardo, R., & Lubell, M. (2016). Understanding What Shapes a Polycentric Governance System. *Public Administration Review, 76*(5), 738–751.

Berkes, F., & Folke, C. (2002). Back to the Future: Ecosystem Dynamics and Local Knowledge. In L. H. Gunderson & C. S. Holling (Eds.), *Panarchy: Understanding Transformations in Human and Natural Systems* (pp. 121–146). Washington, DC: Island Press.

Birkland, T. A. (2005). *An Introduction to the Policy Process*. Armonk, NY: M.E. Sharpe.

Borgatti, S. P., Everett, M. G., & Johnson, J. C. (2013). *Analyzing Social Networks*. London, UK: Sage.

Cairney, P., Fischer, M., & Ingold, K. (2016). Hydraulic Fracturing Policy in the United Kingdom: Coalition, Cooperation, and Opposition in the Face of Uncertainty. In C. M. Weible, T. Heikkila, K. Ingold, & M. Fischer (Eds.), *Policy Debates on Hydraulic Fracturing* (pp. 81–113). Palgrave Macmillan.

Calanni, J. C., Siddiki, S. N., Weible, C. M., & Leach, W. D. (2015). Explaining Coordination in Collaborative Partnerships and Clarifying the Scope of Belief Homophily Hypothesis. *Journal of Public Administration Research and Theory, 25*(3), 901–927.

Crona, B., & Bodin, Ö. (2009). The Role of Social Networks in Natural Resource Governance: What Relational Patterns Make a Difference? *Global Environmental Change, 19*(3), 366–374.

Feiock, R. C., & Scholz, J. T. (2010). *Self-organizing Federalism: Collaborative Mechanisms to Mitigate Institutional Collective Action Dilemmas*. Cambridge, UK: Cambridge University Press.

Fischer, M., & Leifeld, P. (2015). Policy Forums: Why Do They Exist and What Are They Used for? *Policy Sciences, 48*(3), 363–382.

Gerring, J. (2001). *Social Science Methodology: A Critical Approach*. Cambridge, UK: Cambridge University Press.

Hanneman, R. A., & Riddle, M. (2005). *Introduction to Social Network Methods*. Riverside, CA: University of California, Riverside.

Herzog, L. M. (2018). *Cooperation in a Common-Pool Resource Setting: The Case of Micro-Pollutant Regulation in the River Rhine*. Inaugural Dissertation at the Institute of Political Science. Bern, Switzerland: University of Bern.

Herzog, L. M., & Ingold, K. (2019). Threats to Common-Pool Resources and the Importance of Forums: On the Emergence of Cooperation in CPR Problem Settings. *Policy Studies Journal, 47*(1), 77–113.

Huxham, C. (1993). Pursuing Collaborative Advantage. *The Journal of the Operational Research Society, 44*(6), 599–611.

Ingold, K. (2017). How to Create and Preserve Social Capital in Climate Adaptation Polices: A Network Approach to Natural Hazard Prevention. *Ecological Economics, 131*(1), 414–424.

Ingold, K., & Fischer, M. (2014). Drivers of Collaboration to Mitigate Climate Change: An Illustration of Swiss Climate Policy Over 15 Years. *Global Environmental Change, 24*, 88–98.

Ingold, K., & Varone, F. (2012). Treating Policy Brokers Seriously: Evidence from the Climate Policy. *Journal of Public Administration Research and Theory, 22*(2), 319–346.

Jann, W., & Wegerich, K. (2003). Phasenmodelle und Politikprozesse: der Policy Cycle. In K. Schubert & N. C. Bandelow (Eds.), *Lehrbuch der Politikfeldanalyse* (pp. 71–104). München, Germany: De Gruyter.

Jasny, L., & Lubell, M. (2015). Two-Mode Brokerage in Policy Networks. *Social Networks, 41*, 36–47.

Kenis, P., & Schneider, V. (1987). The EC as an International Corporate Actor: Two Case Studies in Economic Diplomacy. *European Journal of Political Research, 15*(3), 437–457.

Knill, C., & Tosun, J. (2012). *Public Policy: A New Introduction.* New York, NY: Palgrave Macmillan.

Knoke, D. (1998). Who Steals My Purse Steals Trash: The Structure of Organizational Influence Reputation. *Journal of Theoretical Politics, 10*(4), 507–530.

Koontz, T. M., & Thomas, C. W. (2006). What Do We Know and Need to Know about the Environmental Outcomes of Collaborative Management? *Public Administration Review, 66*(1), 111–121.

Kümmerer, K. (2009). The Presence of Pharmaceuticals in the Environment Due to the Human Use—Present Knowledge and Future Challenges. *Journal of Environmental Management, 90*, 2354–2366.

Lu, W. B. (2015). Why Do Policy Brokers Matter? A Lesson for Competing Advocacy Coalitions. *Journal of Administrative Sciences and Policy Studies, 3*(2), 33–35.

Lubell, M. (2013). Governing Institutional Complexity: The Ecology of Games Framework. *Policy Studies Journal, 41*(3), 537–559.

Lubell, M., Henry, A. D., & McCoy, M. (2010). Collaborative Institutions in an Ecology of Games. *American Journal of Political Science, 54*(2), 287–300.

Lubell, M., Scholz, J. T., Berardo, R., & Robins, G. (2012). Testing Policy Theory with Statistical Models of Networks. *Policy Studies Journal, 40*, 351–374.

Luxemburger Wort. (2014). Chemische Substanz in der Sauer ist Rapsherbizid: Meazachlor könnte im schlimmsten Fall Folgen für Trinkwasseraufbereitung am Obersauer-Stausee haben. News release, September 29. Retrieved January 26, 2020, from http://www.wort.lu/de/lokales/nach-verunreinigungsvorfall-in-belgien-chemische-substanz-in-der-sauer-ist-rap-sherbizid-542959ceb9b398870806dfbf.

Mayntz, R., & Scharpf, F. W. (1995). *Gesellschaftliche Selbstregelung und Politische Steuerung.* Frankfurt/Main, New York: Campus Verlag.

McGinnis, M. D. (2011). An Introduction to IAD and the Language of the Ostrom Workshop: A Simple Guide to a Complex Framework. *Policy Studies Journal, 39*(1), 169–183.

McGinnis, M. D., & Ostrom, E. (2014). Social-ecological System Framework: Initial Changes and Continuing Challenges. *Ecology and Society, 19*(2), 30.

Metz, F. (2015). *Do Policy Networks Matter to Explain Policy Design? A Comparison of Water Policy Networks and Water Protection Policies for the Reduction of Micropollutants in Four Rhine River Riparian Countries.* Dissertation at the University of Bern, Bern, Switzerland.

Metz, F., & Ingold, K. (2014). Sustainable Wastewater Management: Is It Possible to Regulate Micropollution in the Future by Learning from the Past? A Policy Analysis. *Sustainability, 6*, 1992–2012.

Meyer, B., Pailler, J., Guignard, C., Hoffmann, L., & Krein, A. (2011). Concentrations of Dissolved Herbicides and Pharmaceuticals in a Small River in Luxembourg. *Environmental Monitoring and Assessment, 180*(1–4), 127–146.

Moraes, M. B., Campos, T. M., & Lima, E. (2019). Models of Innovation Development in Small and Median-sized Enterprises of the Aeronautical Sector in Brazil and in Canada. *Gestão and Produção, 26*(1). https://doi.org/10.1590/0104-530X2002-19.

Newig, J., Challies, E., Jager, N. W., Kochskaemper, E., & Adzersen, A. (2018). The Environmental Performance of Participatory and Collaborative Governance: A Framework of Causal Mechanisms. *Policy Studies Journal, 46*(2), 269–297.

O'Leary, R., & Vij, N. (2012). Collaborative Public Management: Where Have We Been and Where Are We Going? *The American Review of Public Administration, 42*(5), 507–522.

Ostrom, E. (1998). A Behavioral Approach to the Rational Choice Theory of Collective Action: Presidential Address, American Political Science Association. *The American Political Science Review, 92*(1), 1–22.

Ostrom, E. (2000). Collective Action and the Evolution of Social Norms. *Journal of Economic Perspectives, 14*(3), 137–158.

Ostrom, E. (2005). *Understanding Institutional Diversity.* Princeton, NJ: Princeton University Press.

Ostrom, E. (2009). A General Framework for Analyzing Sustainability of Social-Ecological Systems. *Science, 325*, 419–422.

Ostrom, E., Gardner, R., & Walker, J. (1994). *Rules, Games and Common Pool Resources.* Michigan, MI: Michigan University Press.

Pal, A., Gin, Y. K., Lin, Y. A., & Reinhard, M. (2010). Impacts of Emerging Organic Contaminants on Freshwater Resources: Review of Recent Occurrences, Sources, Fates and Effects. *Science of the Total Environment, 408*, 6062–6069.

Reffay, C., & Chanier, T. (2003). How Social Network Analysis Can Help to Measure Cohesion in Collaborative Distance-Learning. In *Computer Supported Collaborative Learning* (pp. 343–352, June 2003). Available at https://edutice.archives-ouvertes.fr/edutice-00000422/document.

Sabatier, P. A., & Jenkins-Smith, H. C. (1993). *Policy Change and Learning: An Advocacy Coalition Approach*. Boulder, CO: Westview Press.

Schlager, E. (2004). Common-Pool Resource Theory. In R. F. Durant, D. J. Fiorino, & R. O'Leary (Eds.), *Environmental Governance Reconsidered: Challenges, choices, and opportunities* (American and Comparative Environmental Policy) (pp. 145–175). Cambridge, MA: MIT Press.

Schneider, V. (2014). Akteurskonstellationen und Netzwerke in der Politikentwicklung. In K. Schubert & N. C. Bandelow (Eds.), *Lehrbuch der Politikfeldanalyse* (3rd ed., pp. 259–287). München, Germany: De Gruyter.

Scott, J. (2000). *Social Network Analysis*. London, UK: Sage.

Sohn, C., Reitel, B., & Walther, O. (2009). Cross-border Metropolitan Integration in Europe: The Case of Luxembourg, Basel, and Geneva. *Environment and Planning. C, Government and Policy, 27*(5), 922–939.

Touraud, E., Roig, B., Sumpter, J. P., & Coetsier, C. (2011). Drug Residues and Endocrine Disruptors in Drinking Water: Risk for Humans? *International journal of hygiene and environmental health, 214*(6), 437–441.

Valente, T. (1996). Social Network Thresholds in the Diffusion of Innovation. *Social Networks, 18*, 69–89.

Wasserman, S., & Faust, K. (1994). *Social Network Analysis: Methods and Applications*. Cambridge, UK: Cambridge University Press.

Yla-Anttila, T., Gronow, A., Stoddart, M. C. J., Broadbent, J., Schneider, V., & Tindall, D. B. (2018). Climate Change Policy Networks: Why and How to Compare Them Across Countries. *Energy Research and Social Science, 45*, 258–265.

9

Analyzing Stakeholders' Network to Water Resources Co-management at a Watershed Scale: A Case Study from the Taleghan Watershed in Iran

Fariba Ebrahimiazarkharan, Mehdi Ghorbani, Arash Malekian, and Hans Th. A. Bressers

Introduction

Long before and since the industrial revolution, anthropogenic activities have had negative consequences on water resources (Alilou et al. 2019). A variety of approaches and plans have evolved to assess, evaluate, and monitor their effects. Sustainable development and integrated water resource management (IWRM) have long been considered essential for societal well-being, economic development, and to underpin ecosystem health (Poff et al. 2016). However, researchers have argued that, despite

F. Ebrahimiazarkharan (✉) • M. Ghorbani • A. Malekian
Faculty of Natural Resources, University of Tehran, Tehran, Iran
e-mail: faribaebrahimi@ut.ac.ir; mehghorbani@ut.ac.ir; malekian@ut.ac.ir

H. T. A. Bressers
Department of Governance and Technology for Sustainability (CSTM), University of Twente, Enschede, Netherlands
e-mail: j.t.a.bressers@utwente.nl

© The Author(s) 2020
M. Fischer, K. Ingold (eds.), *Networks in Water Governance*, Palgrave Studies in Water Governance: Policy and Practice, https://doi.org/10.1007/978-3-030-46769-2_9

the global abundance of water and the mainly renewable character of this resource, some people are already, and others will be, living with absolute water scarcity (Hering and Ingold 2012). IWRM can help to coordinate across various goals related to water resources management (Ingold et al. 2016) and a spectrum of issues "which promotes the coordinated development and management of water, land and related resources, in order to maximize the resultant economic and social welfare in an equitable manner without compromising the sustainability of vital ecosystems" (Global Water Partnership 2000).

Until recently, water resource management has been often considered as primarily the exclusive domain of technical experts working under the auspices of the state. Most of their works presumed that water resources could be simulated, predicted, and controlled by means of modeling and infrastructural works. Today, integration and cooperation management have become key principles in IWRM (Newig et al. 2005; Ingold et al. 2016). Water cooperation management has gained increased and wide acceptance (Pahl-Wostl et al. 2007), as has stakeholder involvement in planning and decision-making (Saravanan et al. 2009). Stakeholders can include ministries, resource users/extractors, state agencies, regional governments, NGOs, landowners, and village heads (Bodin and Crona 2009).

Overlapping roles sometimes create diverse, competing interests, and conflict among stakeholders (Blomquist and Schlager 2005). However, co-management that "emphasizes the sharing of rights, responsibilities, and power between different levels and sectors of government and civil society" has helped solve competing interests and conflicts, as well as delivering better IWRM (Huitema et al. 2009). Meanwhile, water governance (WG) has also become an important system covering a range of socio-ecological, political, and economic issues at different levels of society to regulate sustainable water resource management (Rogers and Hall 2003; Gain and Schwab 2012).

Management of water resources in social-ecological systems, such as watersheds, is particularly complex. On the one hand, it has to deal with policy and natural environments that are constantly changing. On the other hand, it involves people who are directly and indirectly influenced by these dynamic environments. Therefore, the management of these systems requires social considerations alongside ecological limits.

In any given watershed, actors at various levels are involved with water governance to achieve sustainable water resource management. Careful analysis of social relationships in community-based networks can identify the key challenges and opportunities. In developing countries, such as Iran, poorly designed water governance structures, a lack of coordination among water sectors, the engagement of a multiplicity of stakeholders and regulators, and hierarchical structures all can exacerbate water crises, as well as contributing to water governance gaps (Madani 2014).

The purpose of our study was to identify key actors in each village and each section of the Taleghan watershed. We used social network analysis (SNA) to increase the awareness of leaders about the power of networks, to expose major relationships, and to strengthen the capacity of the network for cooperation (Hoppe and Reinelt 2010). We assessed issues using indicators at the macro, meso, and micro levels to help to identify means to improve water governance. According to Sell (2016), every sociological theory recognizes a minimum unit (micro) and a maximum unit (macro) with several intermediate levels (meso). Social analysis indicators at each level can reveal metrics to better understand and evaluate leadership networks and key actors (Hoppe and Reinelt 2010). Key actors can offer up ideas, methods, resources, and communication channels to more effectively manage water. Their outcomes as intermediaries can be represented by the extent to which bonding (denotes as connections in a tightly knit group) or bridging occurs (connections to diverse other people) in a network (Putnam 2000; Hoppe and Reinelt 2010). Key actors, who share common interests, can also have a commitment that can influence practice or policy at various levels. These networks are key factors in shaping the social-ecological system (Hoppe and Reinelt 2010). The question is, how effective can they be in delivering sustainable water resource management and governance?

Theory

The literature suggests that restricted water availability for local stakeholders at the watershed scale has created social problems and reduced network cohesion. Identifying the key nodes (actors) that can affect cohesion in a network offer an opportunity for water resource managers to understand who the primary stakeholders are that impact water governance and who are most knowledgeable (Bodin and Prell 2011).

It helps to identify actors involved in water resources management (de Nooy 2010) to facilitate informed decisions on how to pinpoint and select different stakeholders and take advantage of their local knowledge and to involve them in integrated water resource management. Involvement of local leaders and key bridging actors in the knowledge transfer network for water resources management will enhance social capital and social learning, leading to sustainable watershed management.

The most important key local actors are those who play a bridging role, who have high social influence, and who exercise power in their local groups. They use specific mechanisms (e.g. working groups) to link individuals and to create community interaction for management. They can assist regional cooperation and collaboration through their ability to coordinate tasks, build trust, and promote social learning (Kowalski and Jenkins 2015). As such, they can play a key role in solving social-ecological issues by providing expert information and opinion useful to decision-makers (Haas 1992; Kowalski and Jenkins 2015). They can help achieve integrated water resource management by connecting local stakeholders in different social networks and promote group decision-making (Kowalski and Jenkins 2015).

A coherence network can be likened to a spider's web and how connections can act to prevent social conflict between local stakeholders. Key actors amongst network members can solve a range of complex problems in water resources management. A tighter bond between key nodes provides the coherence to reduce weak social ties and promotes certain agents to improve the stability of the social-ecological fit. Any imbalance hinders the achievement of integrated management of natural resources and, thus, sustainable development.

Social network analysis (SNA) has been shown to be a suitable tool to study complex structures and interdependencies, for example, in an IWRM setting. De Nooy (2010) described SNA as a method that "focuses on the structure of ties within a set of social actors, e.g., persons, groups, agencies, and nations, or the products of human activity or cognition such as web sites, semantic concepts, and so on." Stakeholders are pinpointed through applying SNA. Both central and peripheral actors, as well as actors that bridge different parts of the network, are identified in this way. The SNA helps one find key actors amongst stakeholders (Fliervoet et al. 2016). 'Importance' or 'prestige' within a network can be determined in the SNA (Horning et al. 2016).

To achieve efficient structures in IWRM needs understanding of how actors do (and can) manage water resources (de Nooy 2010). SNA also provides guidance for informed decisions on how to identify key bridging actors. Stein et al. (2011) showed how SNA could be applied experimentally to address actors that directly and/or indirectly affected the capacity to govern water in Tanzania. They identified how knowledge from social networks could simplify activities and produce more effective and adaptive IWRM. Similarly, SNA has been used to study complex collaborative relationships between various actors and sectors at multiple levels of flood protection (blue network) and nature management (green network) in the Dutch Rhine delta (Fliervoet et al. 2016). They analyzed how non-governmental actors depended on the main governmental organizations as they played a dominant and controlling role in floodplain management.

Angst and Hirschi (2017) investigated the evolution of a social network of organizational actors involved in the governance of natural resources in a regional nature park project in Switzerland and showed how important were as factors for effective natural resource governance.

Larson et al. (2013) used SNA to explore key stakeholders involved in urban water management. They identified formal and informal interactions and "ideal" networks. They also found that informal networks had created more robust and adaptive urban IWRM.

Özerol et al. (2012) and Özerol and Bressers (2015) showed that sustainable water governance for irrigated agriculture in Turkey was very reliant on the degree of scalar alignment. Horning et al. (2016) used SNA to identify and develop bridging actors in two rural water-scarce regions in Canada to provide enhanced collaboration and increased adaptive capacity within the water governance network.

Material and Methods

Case Study

The study area is located in northern Iran and lies between 50°20′–51°15′E and 36°04′–36°30′N. The Taleghan Watershed drains an area of approximately 948 km² and varies in elevation from 1693 m to 3993 m.a.s.l (Fig. 9.1). Its climate is classified as cold semi-arid, with a mean annual precipitation of 697 mm based on the Köppen-Geiger climate classification system. The dominant precipitation is snow, which generally starts from late fall through to mid-spring.

The Taleghan Watershed is a mountainous area with one main river 'Shahrood.' This is a vital waterway, 105 km long, that supplies much of the drinking water to Tehran and Qazvin city from the Taleghan Dam constructed in 2006. Since then, many changes have occurred, including urban expansion and degradation of the water resources in the study domain. Local stakeholders' social and economic life has been affected by the dam that supports agricultural and horticultural activities. Most of the watershed land cover consists of grassland. Irrigated lands and orchards are the dominant land-use categories along the Shahrood River. Many farmers have sold their agricultural land for residential expansion purposes (e.g. tourist accommodation) at high prices, given the landscape

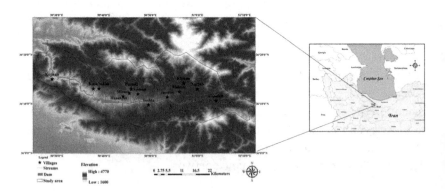

Fig. 9.1 Location of the Taleghan Watershed in Alborz Province, Northern Iran

of the watershed and its view to the dam. Some residents had to emigrate when the dam was filled. This had socio-economic consequences.

Unfortunately, the water flow is insufficient for upstream agricultural land use and also is inadequate for all the downstream users' needs. This has created some conflicts between local actors. Stakeholders are mainly dissatisfied, given the water scarcity for their agricultural activities. Another issue is the reduced water quality and quantity in the study area. These issues affecting conflict between water sectors and stakeholders overlay a lack of a sound governance approach.

Data Collection

Water resource stakeholders (farmers) were identified from information gathered from agriculture organizations in three parts of Taleghan (upstream, center, and downstream of the dam) to identify which villages and who get the most water for their activities. Interviews took place in four rounds covering all seasons (sampling occurred from March 2016 through to December 2017)). This allowed researchers to make estimates about the social network connecting the hidden population (Browne 2005). Snowball and respondent-driven sampling were used to identify 390 local actors. The stakeholders sampled were aged 18–60 years: 150 were from upstream, 150 from the center, and 90 from the downstream areas.

The villages were selected based on parameters including problems with access to water resources, or the permanent presence of more than 50 water users. The upstream area includes 45 villages, wherefrom five villages (Gatedeh, Joestan, Khikan, Mehran, and Narian) with 150 local stakeholders were selected for this study. Similarly, the central area is made up of 25 villages, and 5 villages (Haranj, Hasnjoon, Khasban, Mirash, and Vashteh) with 150 local stakeholders were selected; and the downstream area is made up of 10 villages, and 3 of them (Kajiran, Kash, and Sohan) with 90 local stakeholders were investigated here. The lower number of downstream actors is due to the fact that a lower number of stakeholders is involved in water-related issues: there is a lack of proper access of stakeholders to water, which is why many of them migrated and now the villages are uninhabited.

Information on two key factors—trust and cooperation between stakeholders—was collected by identical questionnaires given out to all actors. Questionnaire design was based on SNA to assess the coherency of collaborative management and to find actors who are able to enhance co-water management.

The outputs (policies and management plans) and outcomes (distribution/user rights among the users) in the three regions of Taleghan watershed can be described as follows: According to the Law on Fair Water Distribution Act (Chap. I), the responsibility for maintaining and permitting and supervising the exploitation of water resources has been delegated to the government and only the beneficiaries can obtain a permit from the Iranian government. Thus, the process of utilization of water resources, both at the decision-making and the implementation level, is entirely at the discretion of the Ministry of Energy (Chap. II of the Fair Distribution Act). Any local level stakeholders, who may play an important role in conservation, are not considered under these laws. Generally, government-oriented system of water resources management excludes local level users in decision-making and implementation.

There exists a completely hierarchical bureaucratic structure for management. This sits alongside a lack of local level authority to adapt management effort to any case-specific circumstances. This increased the gap between lower and upper decision and implementation levels and between downstream and upstream users. This all contributes to inaction at the lower levels regarding rational water allocation, water conservation, and management tasks. This high degree of disconnection occurs across hydrological scales and different topographic locations.

Some rules affect water resources. Upstream, these include gardening permit on steep terrain and licenses to produce medicinal plants and exploit river water resources based on limited water share. In the center, it includes development of sprinkler irrigation and reduced water allocation and water rights for farmers and gardeners. Downstream rules cover permissions to change land use due to a lack of access to sufficient water and abandonment of agricultural land and forced migration of local users of water.

Social Network Analysis (SNA)

At the social networking level, structural indicators were divided into three categories, including indices for the macro level of the network (whole network), meso indicators, and the micro-level (actors) indicators (see Table 9.1) (Wallman 1984; Zhang et al. 2007; Conallin et al. 2017).

Table 9.1 Indicators at the macro/meso/micro levels of the local stakeholder network in Taleghan watershed

Social criteria	Indicators	Level	Ties	Network
Social network cohesion	Density	Macro	Trust and cooperation	Local stakeholder network in water management
Network potential for Exchange Collaboration and Mutual Trust	Transitivity			
Network Potential for Data Transmission and Exchange	Reciprocity			
Identification of networks without exchange group	E-I	Meso		
Determine actors and coherence of central and peripheral networks	Core-periphery			
Identify local stakeholders with a strong reputation and	Degree centrality	Micro		
Local stakeholders have high control and mediation power	Betweenness centrality			

Network Context

Micro-Level Indicators

Micro indexes are indicative of the extent of control an actor has over the flow of information within communication networks. The actor occupying the central position within the network decides whether to disseminate or distort the information he or she receives, whereby to affect the whole network. Actors can be categorized based on network features. These micro indexes range from degree and betweenness centrality (Bonacich 1972; Freeman 1978; Wasserman and Faust 1994; Bodin and Prell 2011), to more elaborate measures of structural equivalence or other positional approaches (Everett and Borgatti 1990; Bodin and Prell 2011).

Centrality is a well-documented indicator to determine the most important actors in a network. The degree of centrality is one of the most important indicators at the level of each actor in determining social power in a network. An actor with a high degree of centrality has more direct connections than other actors (Scott 1988). Betweenness centrality is described as: "the share of times that a node i needs a node k (whose centrality is being measured) to reach a node j via the shortest path" (Borgatti 2005). The assessment reveals gatekeepers and boundary spanners as people who fill structural holes (Palazzolo et al. 2011).

Social power plays a main role in the IWRM. A low degree of centralization also indicates a low level of network cohesion (Bodin and Crona 2009). Key actors have a high level of trust and cooperation in a network and most actors always respect the key actors.

Highly connected individuals in social networks take advantage of their interpersonal ties through friendship or other durable or trustworthy channels to disseminate ideas, information, practices, or resources (Borgatti and Foster 2003; Bodin and Prell 2011). Subgroups differ in the strength of their interactions, but utterly isolated subgroups are rare. The between-group interaction is known as bridging social capital (Woolcock 2001; Bodin and Prell 2011). This is an important condition for engagement of communities with collective action.

For village relations that clearly facilitate transmission of water-related knowledge, this would be regarded as having significant bridging social capital. It is the involvement of the right actors that determines the success of collaborative processes. Who the right actors are is a bone of contention, and this largely depends on the goal of the governance process. Viewed in this way, the identification of stakeholders can be seen as a kind of mobilization of stakeholders.

Findings of SNA can be beneficial for managers or other practitioners to discern those parts of the network which need stronger bridging ties (Prell et al. 2010; Bodin and Prell 2011).

Meso-Level Indicators

Core/periphery structure and E-I indicators were used as meso-level indictors. Palazzolo et al. (2011) showed that a network had a core/periphery structure when "a core whose members are densely tied to each other and a periphery whose members have more ties to core members than to each other." It can help to recognize the presence of a dense, cohesive core and a sparse, unconnected periphery. The E-I indicator model "indicates the extent to which the overall organization is characterized by inter-unit, as opposed to intra-unit, strong ties" (McGrath and Krackhardt 2003). The more negative the E-I indices, the higher the level of informal ties that occurs among people within sub-units.

Thus, an alternative measure of betweenness centrality is suggested that both conceptually and mathematically estimates the number of bridges of an aggregate node (Scott 1988; Bodin and Prell 2011; Bergé et al. 2017). According to Bergé et al. (2017), the number of bridging actors of an aggregate node is a function of three different components: that is, participation intensity (internal); relative outward orientation (external); and diversification of network links (equal). However, these three components are both feasible and attractive to approximate the number of bridges (Bergé et al. 2017). High relative outward orientation refers to the positioning of an aggregate node inclined to use important external knowledge sources. However, a high number of node-internal collaborations would have a negative impact, as it potentially decreases the number of actors connecting different aggregate nodes. In a water resources management context, outward orientation and higher diversification, in particular, may help actors bridging to an aggregate node to develop and renovate their knowledge base faster, or to prevent lock-in situations in certain technologies (Bergé et al. 2017).

Macro-Level Indicators

Macro indicators include coherence, density, transitivity, and reciprocity. These show how the network is structured. Therefore, the effectiveness of the key actors is much higher in cohesive networks in which there are tight connections between actors.

The number of direct links and connections between actors in a network were identified by density. The network could be considered as a coherent network when the number of available links was high. Less dense networks were distinguishable as subgroups and may exert negative effects on the capacity for collaborative processes (Bodin et al. 2006). In contrast, high network density may decrease the groups' effectiveness in collective action (Bodin and Crona 2009), as it can lead to homogenization of knowledge, which decreases the capacity for solving problems (Bodin and Prell 2011).

Transitivity indicates the sustainability of relationships in the network (Coleman 1990; Holling and Meffe 1996). A high number of actors transacting links indicates that sustainability of relationships is strong among actors. However, the stability of the network and the relationship between the relationships were determined by reciprocity. The greater interaction between the actors in the network, the greater stability of the network is shown when the value of reciprocity in the network is high (Wasserman and Faust 1994). Also, it is a good indicator of the mutual involvement of stakeholders. The higher the level of this indicator in the network, the more resilient the social system is.

SNA is applied for measuring and analyzing structured relationships between actors in any water resources context. What makes it extremely precious is its potential for discerning those key actors within the network who make connections between isolated components of the network and also bridge between actors. Actors who occupy such an advantageous position within the network can effect change and they are worthy of more detailed investigation (Burt 2002, 2004, 2005). They can be critical change agents in natural resource governance and pave the way for a move away from unsustainable practices to more sustainable regimes (Olsson et al. 2006; Schultz 2009).

Results and Discussion

Micro-Level Indicators

Results for Degree Centrality

Based on trust and cooperation links, the central actors at the micro level were determined. This dataset was analyzed using UCINET and NetDraw in order to draw the adjacency matrices for better visualization; measure different levels of the structural properties of the network; and analyze the degree centrality to macro-level indexes (Freeman 1978, 2004).

Actors, in this case natural persons using water, located in the center of the network of each village indicated that they have high levels of 'importance' or 'prestige' among the actors (Fig. 9.2). Table 9.2 shows key actors in different parts of the study domain according to their degree centrality. For the stakeholder network in the Taleghan, Lo-Ja (natural person; water user) had a high degree (97%). This indicates a high level of social power in the upstream part with high levels of trust and cooperation and, as a result, the actor can be considered as offering social capital (Falk and Kilpatrick 2000). The same result occurred for the water user Eb-Ga in the downstream part with its degree of 97%. Also, the Md-Shk (97% degree) and Vj-Sn (78% degree) had high levels of social capital (Table 9.2). Therefore, the Lo-Ja, Md-Shk, and Vj-Sn played main roles for IWRM in the upstream part of the Taleghan because they had high levels of trust and cooperation in their network.

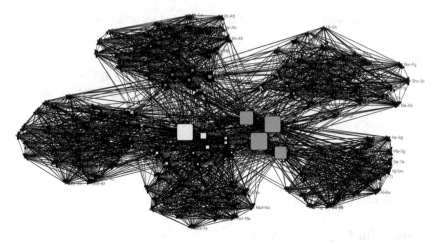

Fig. 9.2 Geometric positioning model of the micro indicators (degree centrality) in the local stakeholders' network in Taleghan water management based on trust/cooperation ties. The nodes' sizes represent the degree centrality. a. Upstream, b. Center, c. Downstream

Table 9.2 Micro indicators (degree and betweenness centralities)

Betweenness centrality		Degree centrality		
Value	Actors	Value	Actors	Area
Linking trust and cooperation (0–100)%				
2.8	Lo-Ja	97	Lo-Ja	Upstream of Taleghan
		95	Sa-Kh	
3	Sa-Kh	93	Ha-Kh	
		78	Vj-Sn	Center of Taleghan
		69	Ha-Ma	
4	Vj-Sn	97	Eb-Ga	Downstream of Taleghan
6	Fa-Mh	96	Md-Shk	
0.4	Eb-Ga	85	Hja-Shk	
0.4	Ya-Al			

Results for Betweenness Centrality

The connection between two actors, who are not connected otherwise, can be linked by a bridge actor (Bodin and Crona 2009). Bridge actors influence information flows and act as mediators or gatekeepers. This is determined here by betweenness centrality (Fliervoet et al. 2016) and the

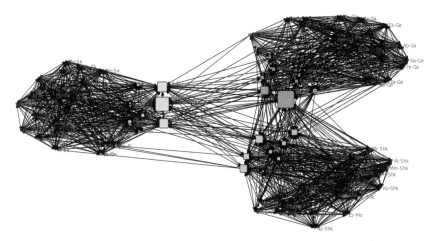

Fig. 9.3 Geometric positioning model of the micro indicators (betweenness centrality) in the local stakeholders' network in Taleghan water management based on trust/cooperation ties. a. Upstream, b. Center, c. Downstream

results are shown in Table 9.2. The actors of the upstream part, including Lo-Ja and Sa-Kh showed the highest betweenness at 2.8 and 3, respectively. In the center part, Vj-Sn and Fa-Mh had high values of 4 and 6, respectively. The value of 0.4 was found for the actors of Eb-Ga and Ya-Al in the downstream part (Fig. 9.3). The figures also indicate that the tendency of actors to communicate between groups in the upstream villages was more than within the other parts. For instance, not many key and bridge actors were present in the downstream network.

Meso-Level Indicators

Results of Core/Periphery Structure

The core/periphery structure shows the presence of a dense, cohesive and sparse, unconnected situation in central and peripheral subgroups (Palazzolo et al. 2011). Table 9.3 shows the cooperation and trust links calculated by the core/periphery index for different subgroups in each part of the study area. The trust and cooperation of actors in the network was 33% in the central subgroups and 66% in the peripheral subgroups.

Table 9.3 Meso indicators (density by core/periphery structure)

Density by periphery structure	Density by core structure	Periphery structure	Core structure	Area
Linking trust and cooperation (0–100)%				
17.5	67.2	66.5	33.5	Upstream
19.8	53.4	83.5	17.2	Center
6.5	45.7	50.5	49.5	Downstream
14.6	55.4	66.8	33.3	Overall

In addition, the density in central subgroups ranged from 45% to 67% and with a low rate of 6% and 17% in the peripheral subgroups. It can be concluded that density in the subgroups was low, indicating that key actors were located in the central subgroups. Therefore, the central group may help us to better understand social power in the network and their potential role in the co-management of water resources. The central group could play the main role in creating opportunities for knowledge exchange and the development of new ideas. This is useful when it comes to making a strong relationship between stakeholders of water resources, especially in IWRM (Horning et al. 2016).

Results of E-I Indictor

Important is also the E-I indicator. Its underlying assumption is that two groups are based on some attribute, one defined as internal and the other as external (Krackhardt and Stern 1988). External links are bridges in the network, which help to exchange more information for their group. This process plays a main role in the success of co-management and organizing activities in a network. On the other hand, a group showing more internal links is more coherent leading to higher levels of trust and sustainable activity in a network (Mohammadi Kangarani et al. 2013). This index ranges from +1 (all links external to subunit) to −1 (all links are internal to subunit). The index is zero when links are divided equally (Krackhardt and Stern 1988). Table 9.4 shows the results of the E-I for the study area. The internal links for the upstream part were high, with a value of +0.41. However, this index was −0.56 and −0.92 for center and downstream areas, respectively. The negative value shows that people are interested in

Table 9.4 Meso indicators (E-I) based on linking trust/cooperation

Density	Possible	Pct	Freq	Area	Parameter
0.80	4350	0.7	3500	Upstream of Taleghan	Internal
0.67	4350	0.78	2916	Center of Taleghan	
0.48	2610	0.96	1274	Downstream of Taleghan	
0.08	2610	0.29	1462	Upstream of Taleghan	External
0.04	1800	0.22	822	Center of Taleghan	
0.09	5400	0.03	50	Downstream of Taleghan	
0.61	13650	−0.41	−2038	Upstream of Taleghan	E-I
0.61	13650	−0.56	−2094	Center of Taleghan	
0.34	2790	−0.92	−1224	Downstream of Taleghan	

enhancing social coherency and in increasing dependency internal to the subunit. The negative E-I indicates all stakeholders tended to more internal links. The upstream stakeholders had less of a tendency to link to the center stakeholders, which was the case for the links between downstream and center stakeholders' network. Hence, this indicates a motivation for all stakeholders of water resources to act selfishly and independently of each other.

Macro-Level Indicators

Results of Density

Density explains the general level of cohesion in a network (Palazzolo et al. 2011). This was a median between all 390 local water stakeholders in the villages (Table 9.2). The upstream social network composed of five villages had a mid-value/level of trust and cooperation ties (50%). The five villages in the center showed overall a 41% level of density in trust and cooperation. In contrast, the three downstream villages showed a low overall level of density in trust and cooperation (38%) in their social network (Table 9.5). Overall, the results show that the density indicator measured was not equal in each region and within subsections of the villages in every region (Bodin and Prell 2011). The upstream stakeholders connected better than the others.

Table 9.5 Macro indicators (density, density overall, transitivity, and reciprocity)

Reciprocity	Transitivity	Density	Density overall	Villages	Area
Linking trust and cooperation (0–100)%					
54.5	58.9	44.4	38.5	Kajiran	Downstream of
59.15	57.2	51.05		Kash	Taleghan
28.45	17.3	21.3		Sohan	
32.05	27.6	42.8	42.6	Haranj	Center of Taleghan
36.6	29.6	50.1		Hasanjon	
26.05	18.5	25.4		Khosban	
38.13	23.4	40.06		Mirash	
38.08	28.7	54.8		Vashteh	
39.05	41.6	58.8	50.6	Gatedeh	Upstream of
24.1	19.25	34.05		Joestan	Taleghan
42.05	43.1	62.9		Khikan	
32.6	25.3	43.9		Mehran	
44.2	32.05	53.6		Narian	

A density of less than 50% is not well connected (Fliervoet et al. 2016). Friedkin (1981) describes a medium density as "a network density of 0.5 indicates that a relation occurs in one-half of the possible pairs in a network." The trust and cooperation among the stakeholders of water resources in the downstream part were at a low level in contrast to elsewhere in the watershed because water resources were in poor condition and water actors had more external relations than in the other parts in Taleghan.

The high level would indicate a positive role in IWRM to help improve the sustainable management of water resources. A low level of density indicates poor co-management of water resources and low coherency of stakeholders (Mohammadi Kangarani et al. 2013). According to this, the coherence of trust network is higher than the cooperation network. Although local stakeholders have trusted each other because of the long historical neighborhood proximity, their participation has diminished by the lack of water resources. It is predicted that scarcity of quality water resources will increase water disputes and, ultimately, not only cohesion but also social trust will be destroyed (the main factor of social capital).

Moreover, the low coherence (40%) was another negative sign in social capital, which can result in water conflicts (Mohammadi Kangarani et al.

2013) and can cause excessive consumption of water resources, especially downstream of the dam.

Results of Transitivity

Transitivity in cooperation networks has been well known as a key index of exclusivity of social networks (Aghagolzadeh et al. 2012). The strength of transitivity index has a direct effect on social cohesion. It also indicates that the sustainability of a network and the relationships among actors are strong (Coleman 1990; Holling and Meffe 1996). The study area's transitivity was at a low level (34%) based on trust and cooperation (Table 9.5). This means that co-management between stakeholders was weak. The levels for upstream, center, and downstream were 32%, 25%, and 44%, respectively. These indicate high sustainability, communications, and bilateral relationships for the networks downstream of the dam. However, the transitivity of communication was less than the average across the whole study area. Therefore, the network has the capability to break down (Coleman 1990; Holling and Meffe 1996). As a result, increased information exchange related to the IWRM requires more education from the water authorities.

Results of Reciprocity

The higher the level of reciprocity in the network, the more resilient is the social system, and the greater the stability of the network (Wasserman and Faust 1994). The 39% reciprocity shows that levels of trust and cooperation were low across the study area (Table 9.5). As was the case for transitivity, there were low levels of co-management and high levels of water conflict between stakeholders (Mostert et al. 2008). The detailed results for each area's trust and cooperation were: 36% (upstream), 34% (center), and 47% (downstream). Again, the resilience and mutual involvement of stakeholders' indicators were higher in the downstream villages.

Conclusion

This study presents local stakeholders' network results for water management in the Taleghan watershed, Alborz, Iran. Water resource management is a main concern of water management organizations operating in this basin. The main purpose was to identify key and bridging actors based on micro indicators (centrality and betweenness degree) in each region who might contribute to and achieve water-compatible governance. Local leaders and key bridging actors in the trust and cooperation network of water resources management can be effective in boosting social capital, as well as cooperation management, and achieving sustainable water resource management in a watershed.

We have tried to define and show how the quantitative social network analysis can help water managers by discovering hidden relationships between water stakeholders and key actors of water resource management. Key actors have strong connections with the rest of local stakeholders in the watershed and they play the most important role in alleviating stakeholders' disputes over limited water resources. They also link water sectors and communicate the demands and suggestions of local stakeholders to managers. Such a connection, in turn, can involve users in local resources management.

Since the construction of Taleghan dam in 2006, access to water resources has decreased significantly, with the quantity and quality of water allocation for local stakeholders in three regions of the basin with different topography. Inappropriate water management at the regional level has increased water conflicts between local stakeholders dramatically. This has affected the socio-ecological system in Taleghan watershed and reduced social cohesion, transitivity, and reciprocity in the local stakeholders' water network. Consequently, the involvement of local stakeholders in management has decreased and management measures have been undertaken ignoring the stakeholders.

The weaknesses of the local stakeholder networks for action plans and co-management of water resources management are shown in Table 9.6.

Measuring macro-level indicators indicated that the cooperation of the network is low compared to trust in the network. The results established

Table 9.6 The weaknesses of the local stakeholders' network

Applied policies	Weakest point	Area	Network	Watershed	Province
Strengthening the links of trust and partnership between individuals	Average cohesion, sustainability, and social capital	Upstream of Taleghan	Local stakeholders in water management	Taleghan	Alborz
	Weak coherence Poor sustainability Low social capital	Center of Taleghan			
	Coherence is very weak Very weak sustainability Low social capital	Downstream of Taleghan			

that the cooperation network upstream (which access to water is naturally more convenient) was stronger than that downstream and in the central section (where the dam was constructed).

Meso-level indicators (E-I indexes) found local water resources stakeholders in all three regions were reluctant to establish intergroup and bridging ties. They were unwilling to link to local stakeholders in other regions of watershed. Low between relations has enhanced water disputes among key and bridge actors across the watershed.

The average cohesion and sustainability and the average social capital are recognized as weaknesses in the upstream network. The central network had the most important weaknesses and showed weak coherence, poor sustainability, and low social capital. Finally, weak coherence, very weak sustainability, and low social capital were characteristics of the downstream local stakeholder networks.

On a wider front, several studies have attempted to create a practical framework to assess social networks. These, and our study, show there needs to be stronger links of theoretical foundations with the actual co-management of water resources. The network analysis approach should be considered as a tool to facilitate socio-ecological system (watershed) management.

Our research results confirm the positive performance of the social network analysis method in identifying key actors. There are many of these amongst the different stakeholders to assist in the policies, planning, and implementation of participatory water resources management and multi-level water governance. This highlights the need to identify, map, and understand different social networks at every level.

This study has provided a novel framework and practical recommendations to detect key water stakeholders at a watershed scale. The study specifically highlights the importance of considering key actors, who are the local decision-makers on water management and who can increase trust, social cohesion, and stakeholder participation in legal or voluntary water management agreements.

References

Aghagolzadeh, M., Barjasteh, I., & Radha, H. (2012). Transitivity Matrix of Social Network Graphs. In *2012 IEEE Statistical Signal Processing Workshop (SSP)* (pp. 145–48). IEEE.

Alilou, H., Nia, A. M., Saravi, M. M., Salajegheh, A., Han, D., & Enayat, B. B. (2019). A Novel Approach for Selecting Sampling Points Locations to River Water Quality Monitoring in Data-scarce Regions. *Journal of Hydrology, 573*, 109–122.

Angst, M., & Hirschi, C. (2017). Network Dynamics in Natural Resource Governance: A case study of Swiss Landscape Management. *Policy Studies Journal, 45*(2), 315–336.

Bergé, L., Scherngell, T., & Wanzenböck, I. (2017). Bridging Centrality as an Indicator to Measure the 'bridging role' of Actors in Networks: An Application to the European Nanotechnology Co-publication Network. *Journal of Informetrics, 11*(4), 1031–1042.

Blomquist, W., & Schlager, E. (2005). Political Pitfalls of Integrated Watershed Management. *Society and Natural Resources, 18*(2), 101–117.

Bodin, Ö., Crona, B., & Ernstson, H. (2006). Social Networks in Natural Resource Management: What is There to Learn from a Structural Perspective? *Ecology and Society, 11*(2).

Bodin, Ö., & Crona, B. I. (2009). The Role of Social Networks in Natural Resource Governance: What Relational Patterns Make a Difference? *Global Environmental Change, 19*(3), 366–374.

Bodin, Ö., & Prell, C. (2011). *Social Networks and Natural Resource Management: Uncovering the Social Fabric of Environmental Governance.* Cambridge, UK: Cambridge University Press.

Bonacich, P. (1972). Factoring and Weighting Approaches to Status Scores and Clique Identification. *Journal of Mathematical Sociology, 2*(1), 113–120.

Borgatti, S. P. (2005). Centrality and Network Flow. *Social Networks, 27*(1), 55–71.

Borgatti, S. P., & Foster, P. C. (2003). The Network Paradigm in Organizational Research: A Review and Typology. *Journal of Management, 29*(6), 991–1013.

Browne, K. (2005). Snowball Sampling: Using Social Networks to Research Non-Heterosexual Women. *International Journal of Social Research Methodology, 8*(1), 47–60.

Burt, R. S. (2002). The Social Capital of Structural Holes. In M. F. Dullen, R. Collins, P. England, & M. Meyer (Eds.), *The New Economic Sociology:*

Developments in an Emerging Field (pp. 148–190). New York, NY: Russel Sage Foundation.

Burt, R. S. (2004). Structural Holes and Good Ideas. *American Journal of Sociology, 110*(2), 349–399.

Burt, R. S. (2005). *Brokerage and Closure: An Introduction to Social Capital.* New York, NY: Oxford University Press.

Coleman, J. S. (1990). *Foundations of Social Theory.* Cambridge, MA: Belknap Press of Harvard University Press.

Conallin, J., Dickens, C., Hearne, D., & Allan, C. (2017). Stakeholder Engagement in Environmental Water Management. In A. C. Horne, J. A. Webb, M. J. Stewardson, B. D. Richter, & M. Acreman (Eds.), *Water for the Environment: From Policy and Science to Implementation and Management* (pp. 129–150). Cambridge, MA: Elsevier.

DeNooy, W. (2010). *Social Network Analysis: Historical Developments and Theoretical Approaches in Sociology* (Vol. 1). Oxford, UK: EOLSS.

Everett, M. G., & Borgatti, S. (1990). A Testing Example for Positional Analysis Techniques. *Social Networks, 12*(3), 253–260.

Falk, I., & Kilpatrick, S. (2000). What is Social Capital? A Study of Interaction in a Rural Community. *Sociologia Ruralis, 40*(1), 87–110.

Fliervoet, J. M., Geerling, G. W., Mostert, E., & Smits, A. J. M. (2016). Analyzing Collaborative Governance Through Social Network Analysis: A Case Study of River Management Along the Waal River in The Netherlands. *Environmental Management, 57*(2), 355–367.

Freeman, L. C. (1978). Centrality in Social Networks Conceptual Clarification. *Social Networks, 1*(3), 215–239.

Freeman, L. (2004). The Development of Social Network Analysis. *A Study in the Sociology of Science, 1,* 687.

Friedkin, N. E. (1981). The Development of Structure in Random Networks: An Analysis of the Effects of Increasing Network Density on Five Measures of Structure. *Social Networks, 3*(1), 41–52.

Gain, A. K., & Schwab, M. (2012). An Assessment of Water Governance Trends: The Case of Bangladesh. *Water Policy, 14*(5), 821–840.

Global Water Partnership Technical Advisory Committee. (2000). *Integrated Water Resources Management.* TAC background paper No. 4. Stockholm, Sweden.

Haas, P. M. (1992). Introduction: Epistemic Communities and International Policy Coordination. *International Organization, 46*(1), 1–35.

Hering, J. G., & Ingold, K. (2012). Water Resources Management: What Should be Integrated? *Science, 336*(6086), 1234–1235.

Holling, C. S., & Meffe, G. K. (1996). Command and Control and the Pathology of Natural Resource Management. *Conservation Biology, 10*(2), 328–337.

Hoppe, B., & Reinelt, C. (2010). Social Network Analysis and the Evaluation of Leadership Networks. *The Leadership Quarterly, 21*(4), 600–619.

Horning, D., Bauer, B. O., & Cohen, S. J. (2016). Missing Bridges: Social Network (dis) Connectivity in Water Governance. *Utilities Policy, 43*, 59–70.

Huitema, D., Mostert, E., Egas, W., Moellenkamp, S., Pahl-Wostl, C., & Yalcin, R. (2009). Adaptive Water Governance: Assessing the Institutional Prescriptions of Adaptive (co-) Management from a Governance Perspective and Defining a Research Agenda. *Ecology and Society, 14*(1), 26.

Ingold, K., Fischer, M., de Boer, C., & Mollinga, P. P. (2016). Water Management Across Borders, Scales and Sectors: Recent Developments and Future Challenges in Water Policy Analysis. *Environmental Policy and Governance, 26*(4), 223–228.

Kowalski, A. A., & Jenkins, L. D. (2015). The Role of Bridging Organizations in Environmental Management: Examining Social Networks in Working Groups. *Ecology and Society, 20*(2).

Krackhardt, D., & Stern, R. N. (1988). Informal Networks and Organizational Crises: An Experimental Simulation. *Social Psychology Quarterly, 51*(2), 123–140.

Larson, S., Alexander, K. S., Djalante, R., & Kirono, D. G. (2013). The Added Value of Understanding Informal Social Networks in an Adaptive Capacity Assessment: Explorations of an Urban Water Management System in Indonesia. *Water Resources Management, 27*(13), 4425–4441.

Madani, K. (2014). Water Management in Iran: What is Causing the Looming Crisis? *Journal of Environmental Studies and Sciences, 4*(4), 315–328.

McGrath, C., & Krackhardt, D. (2003). Network Conditions for Organizational Change. *The Journal of Applied Behavioral Science, 39*(3), 324–336.

MohammadiKangarani, H., Ghonchepour, D., & Holisaz, A. (2013). Social Cooperation Networks and Altered Social Groupings Shaped by Drought: A Case Study of the Village of Tutang in Hormozgon Province. *Desert, 18*(2), 135–144.

Mostert, E., Craps, M., & Pahl-Wostl, C. (2008). Social Learning: The Key to Integrated Water Resources Management? *Water International, 33*(3), 293–304.

Newig, J., Pahl-Wostl, C., & Sigel, K. (2005). The Role of Public Participation in Managing Uncertainty in the Implementation of the Water Framework Directive. *European Environment, 15*(6), 333–343.

Olsson, P., Gunderson, L., Carpenter, S., Ryan, P., Lebel, L., Folke, C., & Holling, C. S. (2006). Shooting the Rapids: Navigating Transitions to Adaptive Governance of Social-Ecological Systems. *Ecology and Society, 11*(1).

Özerol, G., & Bressers, H. (2015). Scalar Alignment and Sustainable Water Governance: The Case of Irrigated Agriculture in Turkey. *Environmental Science & Policy, 45*, 1–10.

Özerol, G., Bressers, H., & Coenen, F. (2012). Irrigated Agriculture and Environmental Sustainability: An Alignment Perspective. *Environmental Science & Policy, 23*, 57–67.

Pahl-Wostl, C., Craps, M., Dewulf, A., Mostert, E., Tabara, D., & Taillieu, T. (2007). Social Learning and Water Resources Management. *Ecology and Society, 12*(2).

Palazzolo, M., Grippa, F., Booth, A., Rechner, S., Bucuvalas, J., & Gloor, P. (2011). Measuring Social Network Structure of Clinical Teams Caring for Patients with Complex Conditions. *Procedia—Social and Behavioral Sciences, 26*, 17–29.

Poff, N. L., Brown, C. M., Grantham, T. E., Matthews, J. H., Palmer, M. A., Spence, C. M., Wilby, R. L., Haasnoot, M., Mendoza, G. F., Dominique, K. C., & Baeza, A. (2016). Sustainable Water Management Under Future Uncertainty With Eco-engineering Decision Scaling. *Nature Climate Change, 6*(1), 25.

Prell, C., Reed, M., Racin, L., & Hubacek, K. (2010). Competing Structure, Competing Views: The Role of Formal and Informal Social Structures in Shaping Stakeholder Perceptions. *Ecology and Society, 15*(4).

Putnam, R. D. (2000). *Bowling Alone: The Collapse and Revival of American Community*. New York, NY: Simon and Schuster.

Rogers, P., & Hall, A. W. (2003). *Effective Water Governance*. TEC Background Papers (No.7). Stockholm: Global Water Partnership.

Saravanan, V. S., McDonald, G. T., & Mollinga, P. P. (2009). Critical Review of Integrated Water Resources Management: Moving Beyond Polarised Discourse. *Natural Resources Forum, 33*, 76–86.

Schultz, L. (2009). *Nurturing Resilience in Social-ecological Systems: Lessons Learned from Bridging Organizations*. Doctoral dissertation, Department of Systems Ecology. Stockholm, Sweden: Stockholm University.

Scott, J. (1988). Social Network Analysis. *Sociology, 22*(1), 109–127.

Sell, C. E. (2016). Max Weber e o átomo da sociologia. Um individualismo metodológico moderado? *Civitas-Revista de Ciências Sociais, 16*(2), 323–347.

Stein, C., Ernstson, H., & Barron, J. (2011). A Social Network Approach to Analyzing Water Governance: The Case of the Mkindo Catchment, Tanzania. *Physics and Chemistry of the Earth, Parts A/B/C, 36*(14–15), 1085–1092.

Wallman, S. (1984). *Eight London Households*. London, UK: Tavistock Publications.

Wasserman, S., & Faust, K. (1994). *Social Network Analysis: Methods and Applications (Structural Analysis in the Social Sciences)*. Cambridge, UK: Cambridge University Press.

Woolcock, M. (2001). The Place of Social Capital in Understanding Social and Economic Outcomes. *Canadian Journal of Policy Research, 2*(1), 11–17.

Zhang, S., Wang, R. S., & Zhang, X. S. (2007). Identification of Overlapping Community Structure in Complex Networks Using fuzzy C-means Clustering. *Physica A: Statistical Mechanics and its Applications, 374*(1), 483–490.

10

Institutional Design and Complexity: Protocol Network Structure in Response to Different Collective-Action Dilemmas

Tomás Olivier, Tyler A. Scott, and Edella Schlager

Introduction

Managing regional natural resources sustainably requires that local governments work together across geographic boundaries and institutional hierarchies. Elected officials, policymakers, and public administrators support coordination and cooperative behavior by devising governing arrangements, encouraging productive interactions, and limiting opportunistic

T. Olivier (✉)
Florida Atlantic University, Boca Raton, FL, USA
e-mail: oliviert@fau.edu

T. A. Scott
University of California, Davis, CA, USA
e-mail: tascott@ucdavis.edu

E. Schlager
University of Arizona, Tucson, AZ, USA
e-mail: schlager@email.arizona.edu

© The Author(s) 2020
M. Fischer, K. Ingold (eds.), *Networks in Water Governance*, Palgrave Studies in Water Governance: Policy and Practice, https://doi.org/10.1007/978-3-030-46769-2_10

behavior. Regional actors often face multiple collective action dilemmas when implementing collaborative agreements because of the different public goods and common pool resources involved (Feiock 2013; Lubell 2013). Thus, governance arrangements are multifaceted, consisting of institutional arrangements such as contracts, rules, and ordinances that provide a context for collective action by mandating and constraining interactions between actors. However, the provision of public goods and the creation of "credible commitments" (Ostrom et al. 1994) require different forms of collective action and, as a result, different types of institutional designs to tackle those problems (Hanlon et al. 2019).

Studies focusing on institutional design for facilitating collective action have mostly developed in parallel to studies of policy or governance networks (hereafter, "governance networks"). The policy networks literature generally portrays a socialized conception of governance networks based on social capital, trust, and other social dynamics. However, social interactions rarely occur in an institutional void. Existing governance network scholarship often overlooks the *protocols* that govern information and resource flows within a network (Galloway and Thacker 2004). Protocols constitute, "the conventional rules and standards that govern relationships within networks" (Galloway and Thacker 2004, p. 8). Particularly in the case of governance networks, protocols shape network structure and function by requiring or incentivizing certain interactions and by specifying how and when interactions take place. This influence can be explicit, such as the case of legally mandated consultation (Amsler 2016), or implicit, such as when rules allocate authority amongst multiple parties who then must work together to address a collective action problem (Feiock 2013). Protocols guide and constrain collective behavior for achieving desired policy and management goals by defining the context within which network interactions occur and by shaping the incentives around actor behavior. By examining the relationships specified by protocols, the structure and topography of governance networks may be better explained.

This chapter focuses on protocols created by formal rules that mandate interactions between a common set of actors to solve different collective action problems. The chapter asks: do protocol networks adopt different structures for the provision of public goods versus the creation of shared decision-making processes and venues? Moreover, are different types of public goods arrangements associated with distinct forms of protocol relations? The social networks literature has shown that network structure is associated with the nature of the dilemma faced by the actors in it (Berardo and Scholz 2010; Angst and Hirschi 2016). We analyze if this also occurs in protocol networks. We draw upon related collective-action literatures to develop hypotheses about expected patterns of interactions. Our hypotheses focus on two factors that might differentiate rule network structures: clustering and connectivity. We associate differences in clustering with rules creating distinct interaction patterns, and connectivity with the extent to which protocol networks rely on multiple or few sets of actors to achieve their collective action goals.

Whereas social network ties are typically modeled as actor-to-actor phenomena, protocols are supra-dyadic (Bonacich et al. 2004), meaning that they specify one-to-one, one-to-many, many-to-one, or many-to-many interactions. We demonstrate alternative methods for analyzing protocol network graphs comprising *hyperedges* connecting multiple actors at once (Butts 2009). We identify three methodological approaches through which these graphs can be analyzed: as *hypergraphs* containing multilateral relations, as *directed bipartite networks* connecting actors and rules, and as *weighted unipartite networks* connecting rules. Since all approaches refer to different network types, we develop and test specific hypotheses for each approach.

Our case is the governing arrangement created to guarantee New York City's drinking water supply. The arrangement creates eleven credible commitment mechanisms (shared decision-making processes and venues) and provide for sixty different public goods, ranging from wastewater treatment plants to job creation programs. We use the rules in this arrangement to devise networks of interactions as prescribed by formal rules (Olivier 2019) around different types of public goods and credible commitments.

Theory

Policy networks are often characterized as consisting of nodes (organizations or actors) and edges (interactions, such as information sharing or participation in joint activities). However, in many policy settings, the nodes and edges that are formed are not freely chosen by the actors, but rather are formed because of protocols. This is readily apparent in the context of governance networks, where regulations, jurisdictional boundaries, property rights, and a host of other formal rules shape many aspects of where, why, and how actors engage with one another.

Scholars across disciplines such as economics (Williamson 1981), political science (Ostrom 1990), public policy (Feiock 2013), and public management (Miller 1992), have focused on explaining how actors resolve collective action problems and provide for shared benefits. In general, two strategies for doing so have been identified: powersharing (Miller 1992) and the delegation of authority (Horn 1995). Powersharing entails actors agreeing to make joint decisions about critical resources. For instance, in the case of the New York City watersheds, New York City and the towns and villages located in the watersheds make joint decisions about watershed lands that are eligible for acquisition by New York City. This replaced a process whereby New York City could condemn and take ownership of lands without consultation as long as a fair market price was paid (NRC 2000). The delegation of authority, on the other hand, involves creating an agency or organization to devise and implement programs, as directed by decision-makers, but insulating the organization from outside interference (Horn 1995). For instance, agency directors, such as in the U.S. Federal Communications Commission or the U.S. Federal Energy Regulatory Commission, often have staggered appointments and longer terms than elected officials, making it difficult to "stack" commissions with supporters.

A key dimension of creating powersharing arrangements or delegating authority is making it very costly for the participants to renege on their commitment to cooperate with one another. Costly reneging is what makes powersharing and authority delegating arrangements credible. These credible commitments constitute the cornerstone on which agreements for the provision or production of public goods are built. For example, if New York City fails to follow the rules mandating it to work with local jurisdictions in acquiring land, the State of New York has the authority to suspend the land acquisition program, which would in turn put the city in noncompliance of federal water quality regulations. If that were to occur, none of the infrastructure or economic development programs would continue. We thus expect rules mandating the creation of powersharing and authority delegation arrangements (credible commitments) to result in network structures that systematically differ from structural arrangements for the production and provision of public goods. Namely, we expect that protocol graphs for credible commitments will evidence stronger clustering—dense patterns of connections between involved parties—and higher overall connectivity.

In turn, public goods can be produced and provided through myriad mechanisms and actors (Oakerson 1999). A robust literature on local public contracting (Savas 1977; Donahue 1989; Stein 1990; Brown and Potoski 2003) has focused on how government protocols adopted to provide for public goods vary by type of good, goals of programs, and market setting. A long running claim in this tradition is that goods that are easily available in markets, such as trash collection or road building, will be contracted for, whereas other types of goods will be provided for through different institutional arrangements (Stein 1990). Building on transaction cost economics (Williamson 1981), recent work has distinguished between *simple* and *complex* public goods, focusing on two features of the contracted-upon good: asset specificity and ease of measurement (Brown and Potoski 2003). Simple public goods have low asset specificity and are easy to measure. Complex public goods, on the other hand, are difficult to measure and their production and provision involves higher transaction costs (Brown et al. 2016). Because of this, actors face higher risks of opportunistic behavior. Protocols providing for complex goods are likely to outline multiple forms and contexts of interactions among the actors involved, in order to hedge against nonperformance. Thus, we expect that complex public good arrangements will cluster more strongly than simple public goods arrangements, closely connecting all involved parties.

Empirical tests of differences in the provision of complex and simple public goods have been mixed (Stein 1990; Hefetz and Warner 2012). As Hefetz and Warner (2012) note, a variety of factors condition how local jurisdictions provide for public goods, besides the nature of the good. In some instances, legislative mandates and policy goals will guide how actors provide for public goods. In other instances, the fiscal capacity and predominant political ideology of a jurisdiction will affect how public goods are provided for (Petersen et al. 2015). For instance, protocols addressing public goods that are critical for the functioning of the agreement or that are more contentious between the parties will receive more attention in their design. We thus expect protocols for salient public goods to tightly constrain the actors' behavior, requiring them to make joint decisions, work closely together, and monitor one another to ensure the goods are being adequately provided for. This translates in an expectation that salient public goods protocols should have higher levels of clustering (indicating rules designed to capture a wider array of scenarios) and connectivity (requiring the actors involved to work closely together).

The literature points out that protocol arrangements should vary in their design, depending on the type of arrangement they create (a credible commitment or a public good arrangement) and on the type of public good to be produced or provided. In turn, we expect those differences to result in variations in clustering and redundancy between protocol networks. We identify three methodological approaches through which these graphs can be analyzed: as *hypergraphs* containing multilateral relations, as *directed bipartite networks* connecting actors and rules, and as *weighted unipartite networks* connecting individual rules. Since tools available for analyzing hypergraphs are not as well developed as those available for analyzing bipartite and unipartite networks, and since measuring techniques vary across two-mode and one-mode networks, we use different network graph measures to compare levels of clustering and connectivity among rule protocols. Doing this requires tailoring our comparisons to reflect the specific indicators of clustering and connectivity in each network type. We develop three sets of hypotheses defining how our concepts of interest manifest in the three different representations of protocol networks: unique subcomponents in protocol hypergraphs, clustering and centralization in protocol bipartite networks, and closeness and redundancy in unipartite rule-to-rule protocol networks.

Because hypergraphs involve more than just pairwise relationships, measures of distance and proximity cannot rely solely on pairwise measures. To assess the differences in structure across the categories of interest, we analyze to what extent hypergraphs can be characterized into distinct subcomponents that reflect relatively isolated components of a graph. Rule networks prescribe different patterns of interactions. If, collectively, rules of a given kind (i.e. simple public goods) are characterized by few graph components, they prescribe patterns of interactions that closely connect all the actors involved. If, on the other hand, rule networks have multiple graph components, then specific responsibilities are assigned to unique individual actors, and are not shared with others. Applying these expectations to subcomponents observed in hypergraphs of network protocols, we posit that:

Hypothesis 1 *Credible Commitments/Complex Public Goods/Salient Public Goods protocols will have fewer* **subcomponents** *than Public Goods/Simple Public Goods/Low-salience Public Goods protocols, respectively.*

Alternatively, bipartite networks provide a different way of assessing protocols, by capturing relationships between rules and the actors mentioned in them. One way of capturing the extent to which rules rely on multiple or fewer actors (as we define protocol connectivity) is to analyze the degree coefficients of actors and rules in bipartite actor-rule networks. Measures of node degree (i.e. the extent of connections a node has) have not been typically used to assess the level of redundancy in a network. However, these measures lend themselves to capture how protocols promote institutional redundancy in the form of rules relying on many or few actors to achieve a certain goal, such as providing for a complex public good. In- and outdegree capture two basic features of institutional design: whether rules are designed to address collective action problems by mandating individual or group action (i.e. whether they rely on many or few actors to provide for a Public Good); and whether actors are assigned redundant responsibilities (i.e. whether they are mandated to send ties—"do something"—multiple times) or instead play few and specific roles overall.

Hypothesis 2a *Credible Commitment/Complex Public Goods/Salient Public Goods protocols will present higher* **rule in- and out-degree** *than Public Goods/Simple Public Goods/Low-salience Public Goods protocols.*

Hypothesis 2b *Credible Commitments/Complex Public Goods/Salient Public Goods protocols will present higher* **actor in- and out-degree** *than Public Goods/Simple Public Goods/Low-salience Public Goods protocols.*

Additionally, bipartite networks allow identifying higher-order measurements to capture redundancy in protocol design. Bipartite clustering coefficients, for instance, provide a higher order measurement of reinforcement between two actors (or two rules) (Opsahl 2013). High levels of clustering indicate that the protocol arrangement requires a high level of redundancy in terms of rule design (rules being interdependent by mandating different actions to the same actors) and in terms of actor behavior (actors required to be involved in the same actions). We thus develop the following expectations for protocols as bipartite networks:

Hypothesis 2c *Credible Commitment/Complex Public Goods/Salient Public Goods protocols will present higher **clustering** than Public Goods/ Simple Public Goods/Low-salience Public Goods protocols.*

Finally, protocol design can also be assessed as a network of interconnected rules. We code a unipartite rule-to-rule network, where two rules are connected if they explicitly mention (and thus aim to influence the behavior of) the same actors. The more compact a rule-to-rule network is, the more connectivity there is among the rules that are part of that specific protocol. By relying on the same actors to achieve their goal (such as creating a credible commitment or providing for a salient public good), protocols create redundant patterns of prescribed interactions. Unlike with actor-rule networks, measures of degree centrality in unipartite protocol networks can only tell whether the protocol has few or many connections between rules, but not necessarily indicate how well connected or close those nodes are. Betweenness and closeness centrality, on the other hand, define how compact (i.e. having multiple, redundant connections) or diffuse (i.e. having few, non-redundant connections) a network is. Rule-to-rule protocol networks with high betweenness have different rules guiding the behavior of distinct actors, and networks with low betweenness have many rules connecting the same actors. High closeness, on the other hand, indicate that a network is compact, whereas low closeness levels indicate that it takes longer paths to reach all other nodes in the network.

Hypothesis 3a *Credible Commitment/Complex Public Goods/Salient Public Goods protocols will present higher* **betweenness centralization** *than Public Goods/Simple Public Goods/Low-salience Public Goods protocols.*

Hypothesis 3b *Credible Commitment/Complex Public Goods/Salient Public Goods protocols will present higher* **closeness centralization** *than Public Goods/Simple Public Goods/Low-salience Public Goods protocols.*

Empirical Setting

The New York City watersheds governing arrangement is the means by which New York City delivers 1.4 billion gallons of largely unfiltered water daily to 9 million people (Finnegan 1997, p. 579). New York City is the largest water provider to operate under a Filtration Avoidance Determination (FAD) granted by the U.S. Environmental Protection Agency. The City sources its water from three watersheds located outside city limits: Croton (10% of the supply), Catskills (40%), and Delaware (50%), covering two thousand square miles. The watersheds are populated with over 100 wastewater treatment plants; hundreds of farms; 128,000 septic tanks, and some manufacturing (Finnegan 1997, p. 581; NRC 2000). The case presents an interesting setting to analyze how a group of government actors (New York City, the watersheds' jurisdictions, New York State, and the U.S. Environmental Protection Agency) jointly developed a highly formal agreement for the governance of a shared natural resource. The governing arrangement provides for shared decision-making and the delegation of authority, as well as for a variety of public goods.

In 1986, the U.S. Congress amended the 1970 Safe Drinking Water Act (SDWA), requiring additional treatment of surface water sources prior to delivery to end users (Gray 1986). In response, the Environmental Protection Agency (EPA) completed a rule specifying criteria that, if not met, would trigger enhanced filtration requirements for surface water sources, known as the Surface Water Treatment Rules (SWTR 1989). If a water utility met the SWTR criteria and demonstrated long-term control over its source watersheds it could apply for a filtration waiver.

In 1992, New York City applied for a filtration waiver (EPA 2002) and the EPA issued a final Filtration Avoidance Determination (FAD) in 1993, lasting until the end of 1996 (EPA 2002).[1] By 1995, in part due to lawsuits filed by watershed jurisdictions over the city's planned acquisition of land in order to meet FAD requirements, concerns arose in the EPA and stakeholders that New York City would not meet its FAD requirements and filtration would be required (EPA 2002). New York Governor Pataki convened the watershed jurisdictions and interested parties to develop a long-term watershed governance approach that resulted in the 1997 Memorandum of Agreement (MOA)—essentially the constitution that created the current polycentric water governance system.

The MOA set the fundamental guidelines for the joint governance of the watersheds, along with a series of water quality protection, infrastructure development, education, and economic development programs to be implemented in the watershed jurisdictions with city funds. In the MOA, the city agreed not to condemn land as its main land acquisition tool and, instead, committed to rely on a "willing-seller, willing-buyer" approach. The MOA also created a conflict resolution venue called Watershed Protection and Partnership Council (WPPC) and a non-profit corporation composed of representatives from the MOA parties, charged with designing and implementing watershed protection programs. This organization is called the Catskills Watershed Corporation (CWC).

Data

We analyze the rules that constitute the governing arrangements of the New York City watersheds as well as the rules that provide for public goods used to meet the goals of the arrangements. The sources of these rules are the 1997 Memorandum of Agreement (MOA), the 2007 New York City Watersheds Rules and Regulations, the 2014 New York City Water Supply Permit, and the Catskill Watershed Corporation

---

[1] New York City disinfects its water using chlorine and ultraviolet disinfection. Beginning in 2015, it began filtering water from the Croton Watershed.

(CWC) program rules. The CWC program rules cover Economic Development, Education, Flood Hazard Mitigation Implementation, Septic Systems, Stormwater Controls, Tax Litigation Avoidance, and Community Wastewater Management Program Rules.

We group rule protocols according to their role in the overall institutional arrangement, such as creating power sharing agreements (credible commitments) or producing/providing for a public good). The MOA created a power sharing agreement by stipulating that land acquisition must be conducted on a "willing seller-willing buyer" basis with participation of the watershed communities involved, and tying this participation to the water supply permit. Credible commitments may also delegate authority to third parties, such as with the creation of the CWC. The CWC was granted the authority to implement watershed protection plans with the approval of representatives from the key signatories to the governing arrangement. In contrast, public good protocols guide the production or provision of a good, such as the construction of wastewater treatment plants or the implementation of economic development plans in the watersheds. Roles were identified by examining each document's article and subsection titles. Next, the content of each subsection was reviewed to determine whether it established a power sharing agreement or, instead, regulated the provision of a public good.

We identified eleven credible commitment arrangements. Sixty public goods arrangements were identified in a similar fashion, and were categorized in two distinct ways. First, each public good was identified as simple or complex. Following the conventions of the local public contracting literature (Lamothe and Lamothe 2012), simple public goods refer to hard infrastructure—roads, bridges, wastewater treatment plants, septic systems—or if outputs are clearly specified and readily measured, such as allocating specific amounts of money to specific jurisdictions (Donahue 1989). Complex public goods are social and economic programs—economic development, education, cultural grants, tax litigation avoidance programs, among others—whose production and outcomes are more difficult to measure. We identified thirty-one simple public goods and twenty-nine complex public goods. Finally, the same public goods were also classified as water quality public goods arrangements and economic development public goods arrangements. Examples of water quality

Table 10.1 Categories of protocols compared (numbers of observations in parenthesis)

Comparison I	Credible commitments (11) vs. Public Goods (60)
Comparison II	Simple public goods (31) vs. Complex Public Goods (29)
Comparison III	Water quality public goods (31) vs. economic development public goods (29)

public goods include programs directly providing water quality services, such as septic systems and wastewater treatment plants. Economic development arrangements, which include main street grants, natural resources grants, and good neighbor payments are classified as less salient for meeting policy goals. Thirty-one water quality and twenty-nine economic development public goods arrangements were identified. Table 10.1 summarizes the categories identified for each comparison.

We inspected each rule that composed a credible commitment or a public good and identified the actors prescribed to interact. For example, the following rule appears in the wastewater treatment program that is part of the New York City watersheds governing arrangements:

If an Identified Community is interested in participating in the [wastewater treatment] program the Community must notify NYCDEP [New York City Department of Environmental Protection], in writing, of its interest no later than July 1, 1997.

The rule connects a Community that is interested in the wastewater treatment program and the New York City Department of Environmental Protection. Rules, or institutional statements, capture an array of instances where actors must perform actions with others, as well as performing actions aimed at inanimate objects (e.g. actor A conducting repairs on a bridge), and actions that have no identifiable recipient (e.g. actor B gathering water quality information). For this analysis, we use only rules that create connections between actors. The number of rules making up a credible commitment or public good ranged from 5 to 96, with an average of 18.83 statements per rule arrangement.

Methods

In analyzing these data, we distinguish between the network—the collection of actors and their various connections as defined by the entire 71 credible commitment and public goods—and a graph, which is the particular set of ties within a given credible commitment or public good. In mathematical terms, a graph is represented by a square matrix where cells take a value of either 0 (no rule linking to actors) or 1 (a rule exists that specifies a relationship between two actors), and the network an array of all 71 graphs that constitute the NYC Watersheds governance arrangement. This formulation differs from many recent network governance studies that analyze only one graph, treating said graph as a stochastic realization of the underlying network, and then analyzing the structural features of the single graph (Scott 2016). This analysis considers multiple graphs that each represent a unique realization of the underlying network to understand whether different structures represent particular collective action challenges.

Protocols as Hypergraph, Bipartite, and Unipartite Networks

Whereas social relationships in policy networks are dyadic (i.e. occur between pairs of individuals or organizations), protocols are *supra-dyadic*, specifying one-to-one, one-to-many, many-to-one, or many-to-many connections (Bonacich et al. 2004).[2] The network science literature refers to a subset of vertices involved in a multilateral relationship as a *hyperedge*, and a graph comprising hyperedges as a *hypergraph* (Bonacich et al. 2004; Butts 2009).

---

[2] It is important to note that, while social relationships are modeled dyadically, a core justification for the use of graph-based modeling approaches is the role of hyperdyadic dependence (i.e. pairwise relationships are influenced by surrounding relationships) (Cranmer and Desmarais 2016). Thus the distinction we draw here is not based on the idea that social dyads are independent (in most cases they are not), but rather that dyads are the base unit in which social ties within governance networks have traditionally been considered (e.g. Desmarais and Cranmer 2012; Yi and Scholz 2016).

Three hyperedges from WSP NYC Watershed Easements agreement

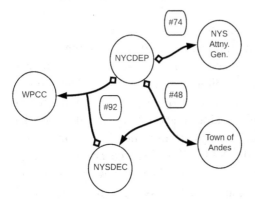

Statement #48: "*The City shall have 10 days from its receipt of the notice to respond in writing to the Contracting Party, the Executive Committee and NYSDEC.*"

Statement #74: "*The City shall provide the Attorney General with reports of all inspections.*"

Statement #92: "*On or before July 1, 1999, the City and State shall each submit, to the WPPC, written progress reports on their respective activities in regard to implementing this Agreement.*"

Fig. 10.1 Example of three hyperedges generated from protocol statements

Figure 10.1 shows an example of three hyperedges created based on institutional statements pertaining to the watershed easements program (a credible commitment protocol). In one case (statement #74), the hyperedge is identical to a normal graph edge, since it simply connects one node to another node; in the other two cases, we see either one party having an obligation to two other parties (statement #48) or two parties having a shared obligation to a third party (statement #92).

Protocols are not the only example of multilateral relationships within governance networks. Other common cases include co-membership in a collaborative group, co-attendance at a meeting, or co-occurrence in media. While the language of hypergraphs and hyperedges is typically not made explicit, this has implicitly been the approach taken in many policy network studies (e.g. Berardo 2014; Ulibarri and Scott 2016; Yi and Scholz 2016; Arnold et al. 2017).

Actors and rules can also be treated as nodes in a bipartite network, but unlike with meeting attendance or co-authorship, ties between actors and rules are directed. Since each protocol specifies something that one or more actors must do to one or more actors, preserving this information requires allowing for directional ties. Using the three hyperedges in Fig. 10.1, Fig. 10.2 shows a bipartite representation of the same data, with statements on one level and actors on another.

Bipartite representation of statements 48, 74, and 92 from WSP NYC Watershed Easements agreement

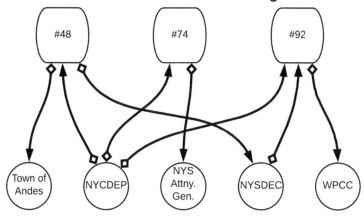

Fig. 10.2 Example of protocol hyperedges converted into bipartite graph format

Although collapsing hyperedges into a unimodal projection can cloud the underlying multilateral structure of the data, we believe this is more of a concern when considering actor-to-actor connections in a protocol network, since single rules often connect multiple actors. In contrast, a useful alternative is to analyze a rule-to-rule projection of the bipartite network, which can help assess the extent to which a governance arrangement (i.e. a credible commitment or a public good) involves many rules connecting the same actors versus different rules connecting different actors. Further, because a pair of rules can have multiple shared actors, the projection is weighted—that is, comprising integer values reflecting the number of shared actors between two rules. By transposing the actor-rule bipartite networks into a rule-rule network, we present a third approach to capture how close or loosely coupled are the institutional statements that constitute a credible commitment or public good rule protocol.

The rule-to-rule network captures connections between institutional statements within a protocol, where a connection exists if two rules guide the same actor's behavior and where those connections are weighted, that is, the more actors are shared by two rules, the "stronger" their connection will be. In our case, the unimodal graph is undirected. For instance, in the example shown in Fig. 10.2, the transpose now shows three edges: #48—#74, #74—#92, and #48—#92.

Subcomponents and Clustering in Actor-Actor Hypergraphs

One increasingly deployed method of partitioning hypergraphs is spectral clustering (Agarwal et al. 2005; Zhou et al. 2007; Li et al. 2014). Spectral clustering proceeds in three basic steps: (1) represent graph vertices in a similarity matrix; (2) compute the graph Laplacian, or the degree matrix subtracted by the adjacency matrix; (3) find (in our case) 2 eigenvalues and corresponding eigenvectors for the graph Laplacian; and (4) applying k-means clustering to the eigenvectors (each set of eigenvectors representing a graph vertex) to determine an optimal number of partitions (Hastie et al. 2005; Kawa 2018). We conduct this process in R adapting code generated by Kawa (2018). For each graph, we apply the process described above and then, in step 4, test a range of possible k values (k represents the number of clusters) and identify the optimal k value for each graph using the *NBClust* package in R (Charrad et al. 2014).

Connectivity and Clustering in Actor-Rule Bipartite Networks

For each bipartite graph consisting of actors and rules or protocols, we compute basic summary statistics reflecting overarching structure differences. We focus on three basic structural concepts that reflect connectivity in each graph:

1. Average indegree for actors and for rules.
2. Average outdegree for actors and for rules.
3. A clustering coefficient for bipartite graphs which measures the probability that a bipartite three-path (e.g. from $actor_i \rightarrow rule_1 \rightarrow actor_j$ $\rightarrow rule_2$) is part of a bipartite four-cycle (i.e. $actor_i \rightarrow rule_1 \rightarrow actor_j$ $\rightarrow rule_2 \rightarrow actor_i$) (Robins and Alexander 2004). This measure is conceptualized as a measure of reinforcement between two actors (or two rules) (Opsahl 2013).

Connectivity in Rule-Rule Networks

The primary measure of rule connectivity we use is the weighted betweenness and closeness measures developed by Opsahl (2009; see also Opsahl and Panzarasa 2009). Network closeness is the distance in path length to other nodes, and betweenness is the extent to which a node exists on the shortest path between other nodes in the network (Freeman 1979). The measures we use are generalizations to weighted graphs which reflect where larger edge weights (e.g. two rules which apply to more of the same actors) are more strongly connected (Newman 2001). Both measures are calculated at the nodal level and then we compute average betweenness and closeness values for each graph.

Results

Spectral Clustering of Protocol Hypergraphs

Hypothesis 1 associates the presence of multiple patterns of interactions (clusters) with different rule strategies aimed at dealing with specific issues and argues that credible commitment, complex public goods, and salient public goods (water quality) protocols should present fewer clusters than their counterparts. A hypergraph with one cluster indicates that there is one component created by all the institutional statements in that rule protocol. Consequently, more clusters are associated with different subcomponents, and thus with rules assigning unique responsibilities to individual actors.

Figure 10.3 presents the optimal number of clusters identified for hypergraphs of credible commitments and for public goods. The results show that while the average number of optimal clusters is slightly higher for public goods, the difference is minor and not statistically significant. Thus, we find no support for hypothesis 1 with respect to clustering in public goods versus credible committment arrangements.

As shown in Fig. 10.3, the majority of credible commitment and public good arrangements share a common three-cluster structure. This

# of subcomponents in institutional arrangement graphs

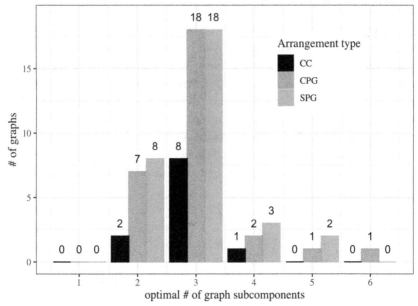

Fig. 10.3 Subcomponents in credible commitment, simple public good, and complex public good hypergraphs

Table 10.2 Hypergraph subcomponents in credible commitments and public good arrangements

Type	N	Avg. subcomponents	Std. Dev subcomponents
Credible commitment	11	2.91	0.54
Public good	60	2.98	0.83

means that system actors tend to develop similar high-level configurations for dealing with a variety of collective action problems involved in the provision and production of public goods. The comparisons between public good types (comparisons II and III) yielded no salient differences either. In all cases, the mode was three subcomponents, with no major substantive differences between public good types.

Redundancy in Bipartite Actor-Rule Networks

The analysis of bipartite actor-rule networks yields results similar to the analysis of hypergraphs. Credible commitment protocols tend to be designed with higher values of rule indegree, actor indegree, actor outdegree, and clustering coefficients than public goods (Table 10.3). However, Welch's t-test comparisons (an adaptation of the standard Student's t-test for comparing two samples with different sample sizes and/or different variances) (Derrick et al. 2016) between the two groups only found statistically significant differences in clustering coefficients ($p > 0.05$). This indicates that credible commitments are deliberately designed to include more redundancy (in the form of higher clustering) when compared to public goods (Hypothesis 2c).

Regarding comparisons II and III, the analysis provided no support for our hypotheses. We encountered statistically significant differences in the average rule indegree between economic development and water quality ($p < 0.05$), against Hypothesis 2a. With regards to the comparison between complex and simple public goods, we observed statistically significant differences in average actor indegree ($p < 0.1$) and bipartite clustering coefficients. Simple public goods present higher average actor indegree and higher bipartite clustering coefficient, thus not supporting Hypotheses 2b and 2c for this comparison.

Connectivity in Rule-to-Rule Networks

Table 10.4 presents results for the three comparisons in unipartite rule-to-rule networks. We again used Welch's t-tests to assess differences between the groups of interest. Results showed statistically significant results for the differences on average rule betweenness ($p < 0.05$) and average rule closeness ($p < 0.1$) between credible commitments and public goods, only providing support for Hypothesis 3b and failing to support Hypothesis 3a. Higher average closeness indicates that credible commitment rule networks rely on many institutional statements connecting the same actors, creating complementarity between individual rules. Lower average betweenness indicate that credible commitment networks are less dependent on a few rules to keep the network together than

Table 10.3 Comparison between bipartite actor-rule protocol networks

| Comparison | Type | N | Hypothesis 2a | | Hypothesis 2b | | Hypothesis 2c |
			Avg. rule indegree	Avg. rule outdegree	Avg. actor indegree	Avg. actor outdegree	Clustering coef. (undirected)
I	Credible commitment	11	12.117	1.026	0.329	3.226	0.128**
	Public good	60	11.427	1.197	0.195	1.548	0.051**
II	Complex public good	29	10.898	1.018	0.153*	1.36	0.035*
	Simple public good	31	11.922	1.365	0.233*	1.725	0.066*
III	Economic development	31	13.383**	1.226	0.17	1.761	0.063
	Water quality	29	9.336**	1.166	0.221	1.321	0.039

*p < 0.10; **p < 0.05

Table 10.4 Comparison between credible commitments and public goods rule-to-rule networks

Comparison	Type	N	Hypothesis 3a Average betweenness	Hypothesis 3b Average closeness
I	Credible commitment	11	234.926**	13.018*
	Public good	60	322.527**	9.422*
II	Complex public good	29	267.201*	8.209**
	Simple public good	31	374.283*	10.556**
III	Economic development	31	337.847	9.849
	Water quality	29	306.149	8.965

$^*p < 0.10$; $^{**}p < 0.05$

public good networks. Results also showed statistically significant differences in the average betweenness centralization and average closeness between complex and simple public goods. In both cases, however, the results failed to support Hypotheses 3a and 3b.

Discussion

Our analysis shows that actors purposefully design rules according to the type of collective-action problem they intend to address. However, this difference is only noticeable for rules designed to provide the foundations of a governing arrangement, via powersharing and delegation of authority, when compared to those defining how the arrangement will produce or provide for public goods (comparison I). Results provide support for some of our hypotheses in the comparison between credible commitments and public goods. Credible commitment rule protocols tend to be more customized than public good protocols, as shown by the higher clustering coefficients in bipartite actor-rule networks, and the clustered and compact nature of their unipartite networks. These findings coincide with the expectation that providing credible commitments require mechanisms that "tie the actors' hands together," which is partially achieved through highly customized rules that anticipate myriad scenarios of opportunistic behavior and through the creation of multiple and redundant responsibilities among the actors involved. The New York City watershed arrangements both (a) include clear and explicit credible commitment arrangements and (b) design such arrangements in a unique fashion.

Protocol networks of public goods, on the other hand, do not present consistent differences in their design. None of our hypotheses were supported across the public good comparisons (comparisons II and III). The only major differences were both type- and network-specific. For instance, bipartite actor-rule networks of water quality presented significantly lower rule-indegrees than their economic development counterparts. This finding indicates that water quality rules tend to assign fewer actors in charge of conducting an action than economic development rules, which we interpret as indicating that water quality rules are more targeted than economic development rules. Water quality arrangements, as salient public goods in the New York City Watersheds agreement, were expected to be more redundant (by mandating more joint decisions and work closely). Lower rule indegrees may indicate that the opposite is happening. However, since the maintenance of water quality is relatively easy to target (identifying sources of pollution, retrofitting infrastructure, developing best management practices, among others), it could be that "more constraint" in this context indicates highly targeted rules. This context-specific interpretation of our expectations may also explain the unexpected results observed in Hypotheses 3a and 3b for this comparison.

Overall, these unexpected findings in the comparisons between types of public goods can be attributed to two sources: first, the expectations from the contracting and public good provision literatures may require more fine-tuning, adjusting them both to the specific networks studied (hypergraph, bipartite, or unipartite), and to the nature of the public good being analyzed (low rule indegree in a bipartite network may indicate very different processes in water quality than in simple or complex public goods). Second, the literature studying policy network structure and their relation to collective-action outcomes is still under development. Perhaps the differences between public good types are more subtle than just the presence of redundancy or more clusters, and, instead, more specific indicators may be required for these comparisons.

Conclusion

In this chapter, we have analyzed the design of a regional governing arrangement for the provision of unfiltered drinking water, applying concepts from network theory and social network analysis. We studied whether rule networks created to address different types of collective action problems present differences in their clustering and connectivity. Formal rules, just like networks, are relational in nature, and are only effective in the context that they can clearly indicate behavioral expectations and establish mechanisms through which those expectations are enforced. Understanding how formal rules establish such expectations, and how they vary in specific collective-action scenarios is critical for identifying the mechanisms through which actors can overcome collective action problems.

By analyzing the over 3000 rules that constitute the New York City Watersheds regional governing arrangement, we identified how rules define behavioral expectations in terms of relations between the parties to the agreement. Applying recent techniques to "translate" rules into meaningful networks, we assessed whether patterns of rule-prescribed interactions varied when actors are trying to address different kind of collective action problems. Unlike traditional social network analyses, where patterns of interactions are created as an aggregation of dyadic interactions, rule-prescribed networks are supra-dyadic in that a single rule can create a one-to-many relationship. Conceptually, this is different than aggregating many one-to-one relationships to recreate that similar pattern. Acknowledging these conceptual nuances, we applied three types of analyses to study our rule networks: we studied them as hypergraphs, as bipartite actor-to-rule-to-actor networks, and as unipartite rule-to-rule networks.

Findings showed systematic differences between rule protocols designed to create powersharing and delegation commitments and rule protocols created to produce and provide public goods, when analyzed as bipartite actor-rule networks and as unipartite rule-rule networks. These align with theory expectations, which highlight the critical role of making exit and opportunistic behavior costly to create robust institutional arrangements. Aside from a few specific instances, our analyses failed to support our expectations in the comparisons between types of public good arrangements across hypergraph, bipartite, and unipartite rule-based networks.

This work constitutes one of the first attempts to bridge the gap between social network studies and studies on institutional design. However, future research should address some of our limitations. First, our chapter focused only on a within-case comparison (the New York City Watersheds case). Future research should analyze whether other institutional arrangements also design credible commitments in similar ways, comparing arrangements at different scales (i.e. international agreements; multilateral and bilateral agreements). Second, our analysis is static. By incorporating a dynamic component of rule changes over time, our methodological approach can help provide new insights for the study of institutional change. Finally, while protocols incentivize and constrain interactions, it remains to be seen how protocol-bound actors interact in reality—that is, what types of relational behavior unfold within different protocol arrangements? Future studies that incorporate both protocol structures and socio-relational data will be able to more fully understand the structure and function of complex governance networks.

References

Agarwal, S., Lim, J., Zelnik-Manor, L., Perona, P., Kriegman, D., & Belongie, S. (2005). Beyond Pairwise Clustering. In *2005 IEEE Computer Society Conference on Computer Vision and Pattern Recognition*. CVPR 2005 (Vol. 2, pp. 838–845).

Amsler, L. B. (2016). Collaborative Governance: Integrating Management, Politics, and Law. *Public Administration Review, 76*(5), 700–711.

Angst, M., & Hirschi, C. (2016). Network Dynamics in Natural Resource Governance: A Case Study of Swiss Landscape Management. *Policy Studies Journal, 45*, 315–336.

Arnold, G., Nguyen Long, L. A., & Gottlieb, M. (2017). Social Networks and Policy Entrepreneurship: How Relationships Shape Municipal Decision Making About High-Volume Hydraulic Fracturing. *Policy Studies Journal, 45*(3), 414–441.

Berardo, R. (2014). Bridging and Bonding Capital in Two-Mode Collaboration Networks. *Policy Studies Journal, 42*(2), 197–225.

Berardo, R., & Scholz, J. T. (2010). Self-Organizing Policy Networks: Risk, Partner Selection, and Cooperation in Estuaries. *American Journal of Political Science, 54*(3), 632–649.

Bonacich, P., Holdren, A. C., & Johnston, M. (2004). Hyper-Edges and Multidimensional Centrality. *Social Networks, 26*(3), 189–203.

Brown, T. L., & Potoski, M. (2003). Contract-Management Capacity in Municipal and County Governments. *Public Administration Review, 63*(2), 156–164.

Brown, T. L., Potoski, M., & Van Slyke, D. (2016). Managing Complex Contracts: A Theoretical Approach. *Journal of Public Administration Research and Theory, 26*(2), 294–308.

Butts, C. T. (2009). Revisiting the foundations of network analysis. *Science, 325*(5939), 414–416.

Charrad, M., Ghazzali, N., Boiteau, V., & Niknafs, A. (2014). NbClust: An R Package for Determining the Relevant Number of Clusters in a Data Set. *Journal of Statistical Software, 61*(4), 1–36.

Cranmer, S. J., & Desmarais, B. A. (2016). A Critique of Dyadic Design. *International Studies Quarterly, 60*(2), 355–362.

Derrick, B., Toher, D., & White, P. (2016). Why Welch's Test Is Type I Error Robust. *Tutorials in Quantitative Methods for Psychology, 12*(1), 30–38.

Desmarais, B. A., & Cranmer, S. J. (2012). Micro-Level Interpretation of Exponential Random Graph Models with Application to Estuary Networks. *Policy Studies Journal, 40*(3), 402–434.

Donahue, J. (1989). *The Privatization Decision: Public Ends, Private Means.* New York, NY: Basic Books.

Feiock, R. C. (2013). The Institutional Collective Action Framework. *Policy Studies Journal, 41*(3), 397–425.

Finnegan, M. C. (1997). New York City's Watershed Agreement: A Lesson in Sharing Responsibility. *Pace Environmental Law Review, 14*(2), 577–644.

Freeman, L. C. (1979). Centrality in Social Networks Conceptual Clarification. *Social Networks, 1*(3), 215–239.

Galloway, A. R., & Thacker, E. (2004). Protocol, Control, and Networks. *Grey Room, 17*, 6–29.

Gray, K. F. (1986). The Safe Drinking Water Act Amendments of 1986: Now a Tougher Act to Follow. *Environmental Law Reporter, 16*, 10338–10345.

Hanlon, J. W., Olivier, T., & Schlager, E. (2019). Suspicious Collaborators: How Governments in Polycentric Systems Monitor Behavior and Enforce Public Good Provision Rules Against One Another. *International Journal of the Commons, 13*(2), 977–922.

Hastie, T., Tibshirani, R., Friedman, J., & Franklin, J. (2005). The Elements of Statistical Learning: Data Mining, Inference and Prediction. *The Mathematical Intelligencer, 27*(2), 83–85.

Hefetz, A., & Warner, M. (2012). Contracting or Public Delivery? *Journal of Public Administration Research and Theory, 22*(2), 289–317.

Horn, M. (1995). *The Political Economy of Public Administration: Institutional Choice in the Public Sector*. New York, NY: Cambridge University Press.

Kawa, N. (2018). *Spectral Clustering*. (Online). Retrieved from https://rpubs.com/nurakawa/spectral-clustering.

Lamothe, M., & Lamothe, S. (2012). What Determines the Formal Versus Relational Nature of Local Government Contracting? *Urban Affairs Review, 48*(3), 322–353.

Li, X., Hu, W., Shen, C., Dick, A., & Zhang, Z. (2014). Context-Aware Hypergraph Construction for Robust Spectral Clustering. *IEEE Transactions on Knowledge and Data Engineering, 26*(10), 2588–2597.

Lubell, M. (2013). Governing Institutional Complexity: The Ecology of Games Framework. *Policy Studies Journal, 41*(3), 537–559.

Miller, G. (1992). *Managerial Dilemmas: The Political Economy of Hierarchy*. New York, NY: Cambridge University Press.

National Research Council. (2000). *Watershed Management for Potable Water Supply: Assessing the New York City Strategy*. Washington, DC: National Academies Press.

Newman, M. E. J. (2001). Scientific collaboration networks. II. Shortest paths, weighted networks, and centrality. *Physical Review. E, Statistical, Nonlinear, and Soft Matter Physic, 64*, 016132.

Oakerson, R. (1999). *Governing Local Public Economies: Creating the Civic Metropolis*. Oakland, CA: ICS Press.

Olivier, T. (2019). How Do Institutions Address Collective-Action Problems? Bridging and Bonding in Institutional Design. *Political Research Quarterly, 72*(1), 162–176.

Opsahl, T. (2009). *Structure and Evolution of Weighted Networks*. London: University of London (Queen Mary College).

Opsahl, T. (2013). Triadic Closure in Two-Mode Networks: Redefining the Global and Local Clustering Coefficients. *Social networks, 35*(2), 159–167.

Opsahl, T., & Panzarasa, P. (2009). Clustering in Weighted Networks. *Social networks, 31*(2), 155–163.

Ostrom, E. (1990). *Governing the Commons: The Evolution of Institutions for Collective Action.* New York, NY: Cambridge University Press.

Ostrom, E., Gardner, R., & James, W. (1994). *Rules, Games, and Common-Pool Resources.* Michigan, MI: University of Michigan Press.

Petersen, O. H., Houlberg, K., & Christensen, L. R. (2015). Contracting Out Local Services: A Tale of Technical and Social Services. *Public Administration Review, 75*(4), 560–570.

Robins, G., & Alexander, M. (2004). Small Worlds Among Interlocking Directors: Network Structure and Distance in Bipartite Graphs. *Computational and Mathematical Organization Theory, 10*(1), 69–94.

Savas, E. S. (1977). An Empirical Study of Competition in Municipal Service Delivery. *Public Administration Review, 37*(6), 717–724.

Scott, T. A. (2016). Analyzing Policy Networks Using Valued Exponential Random Graph Models: Do Government-Sponsored Collaborative Groups Enhance Organizational Networks? *Policy Studies Journal, 44*(2), 215–244.

Stein, R. (1990). *Urban Alternatives: Public and Private Markets in the Provision of Public Services.* Pittsburgh, PA: University of Pittsburgh Press.

Surface Water Treatment Rule, 40 C.F.R §141.71. (1989).

Ulibarri, N., & Scott, T. A. (2016). Linking Network Structure to Collaborative Governance. *Journal of Public Administration Research, 27*(1), 163–181.

United States Environmental Protection Agency. (2002). *New York City Filtration Avoidance Determination.* Retrieved from https://archive.epa.gov/region02/water/nycshed/web/pdf/2002fad.pdf.

Williamson, O. E. (1981). The Economics of Organization: The Transaction Cost Approach. *American Journal of Sociology, 87*, 548–577.

Yi, H., & Scholz, J. T. (2016). Policy Networks in Complex Governance Subsystems: Observing and Comparing Hyperlink, Media, and Partnership Networks. *Policy Studies Journal, 44*(3), 248–279.

Zhou, D., Huang, J., & Schölkopf, B. (2007). Learning with Hypergraphs: Clustering, Classification, and Embedding. In B. Schölkopf, J. C. Platt, & T. Hoffman (Eds.), *Advances in Neural Information Processing Systems 19* (pp. 1601–1608). Vancouver: MIT Press.

11

Comparing Centrality Across Policy Networks and Media Narratives

Emily V. Bell and Adam Douglas Henry

Introduction

This chapter considers two of the most widely used measures of actor position in network analysis—degree centrality and betweenness centrality—and considers the extent to which these measures capture the theoretical concepts of policy entrepreneurship and brokerage. While network analysis is widely applied to the study of policy systems, researchers often assume that certain network metrics adequately capture particular theoretical concepts. These assumptions may have face validity, but validation is important for advancing the utility of network analysis to policy theory. We work towards this goal by comparing empirical, quantitative

E. V. Bell (✉)
University of Georgia, Athens, GA, USA
e-mail: emily.bell@duke.edu

A. D. Henry
University of Arizona, Tucson, AZ, USA
e-mail: adhenry@email.arizona.edu

© The Author(s) 2020   **295**
M. Fischer, K. Ingold (eds.), *Networks in Water Governance*, Palgrave Studies in Water
Governance: Policy and Practice, https://doi.org/10.1007/978-3-030-46769-2_11

measures of network position with narratives of actor importance that we measure through qualitative analysis of media narratives.

Our empirical context is local water governance in a single policy subsystem in Tucson, Arizona, a semi-arid desert municipality in the southwestern United States. Systematic analysis of media reports and gray literature is used to identify policy innovations surrounding water sustainability across multiple domains such as stormwater management, water quality, and water supply. We also rely on this literature to identify actors participating in these innovations, yielding a two-mode network of actors to policy events, from which a unipartite network of actor-actor connections is created. Centrality in this network is correlated with narrative accounts of the importance of these actors, as well as the narrative classification of these actors as policy brokers and policy entrepreneurs. Results show that network centrality measures do correlate well with media accounts of importance in the policy system, a finding which provides a stronger linkage between frequently used empirical network measures and key theoretical concepts in policy theory.

After we present a brief overview of our case, we discuss our data collection process and then assess the similarity of reports with network measures. We conclude by reviewing our findings and present considerations for using content analysis and network analytic techniques for validation and broader lessons for future research.

Theory: Network Centrality and Actor Importance

Research on the policy process often focuses on the role of brokers and entrepreneurs in policy networks. These types of actors are often described as well-embedded in policy systems, with the ability to navigate competition among policy stakeholders, and the power to influence policy outcomes (Sabatier and Jenkins-Smith 1999). These roles depend upon developing relationships to other policy network participants, and thus the concepts of policy brokerage and entrepreneurship are commonly operationalized in network terms.

Policy Brokers

Brokers link otherwise sparsely or disconnected groups of actors in order to attenuate conflict and to foster communication (Sabatier and Jenkins-Smith 1993). In policy settings characterized by competing coalitions, those in brokerage roles can mediate conflict to achieve compromise, enabling stakeholders to arrive at policy outcomes that receive shared support (Sabatier and Jenkins-Smith 1999). Empirical research has shown that brokerage can enhance policy outcomes in a variety of policy settings, such as navigating new flood risk management plans that incorporate integrated water resource management approaches (Rouillard et al. 2013), or retrofitting urban settings to restore balance between natural and human systems (Karvonen 2010). New information from this process and novel policy outcomes can also lead coalitions to update their policy-core beliefs in a process commonly referred to as *policy-oriented learning* (Sabatier and Jenkins-Smith 1993). Learning is hypothesized to have a connection to policy change, although empirical evidence of this linkage has been varied (Weible et al. 2009).

Policy brokers can also connect network participants that are otherwise segregated by functional specialization, responsibility, or scale for sharing knowledge and resources (Ernstson et al. 2010; Sharma et al. 2012; Feiock 2013). Brokerage in low-conflict settings is shown to consolidate and process complex information from multiple sources (e.g., local and expert knowledge) for stakeholders involved in natural disaster response and resilience-building efforts (Tompkins et al. 2002; Olsson et al. 2007; Sitas et al. 2016).

Policy Entrepreneurs

Like policy brokers, policy entrepreneurs are well connected, but also seek to mobilize support and to access novel and timely resources necessary for advancing policy alternatives to the agenda (Mintrom and Vergari 1998). The concept of the policy entrepreneur emerged from the agenda-setting literature, which suggests that these actors identify and frame current issues, draw the attention of stakeholders, and seize "windows of opportunity" to present policy problems and solutions to decision makers (Weible 2007; Kingdon 1995). To fulfill these tasks, policy entrepreneurs seek novel information in the policy network to stay alert to emergent opportunities and to know when to mobilize others to achieve policy objectives (Christopoulos and Ingold 2015).

The urban water governance literature shows that policy entrepreneurs facilitate change by stimulating momentum for the adoption and implementation of new water management strategies (Brown 2008; Bettini et al. 2015). This behavior can promote transitions to adaptive water management strategies, such as the use of integrated technologies to improve efficiency between different water service areas (Keremane 2015) and other techniques that are designed to improve resilience to climate uncertainty (Brouwer and Biermann 2011). Finally, where water professionals and decision makers are averse to risk and experimentation (Farrelly and Brown 2011), policy entrepreneurs can demonstrate the feasibility of proposed policy solutions (Mintrom and Norman 2009).

Policy entrepreneurs are often described as "leaders" in driving policy change (Gunderson and Light 2006; Olsson et al. 2006; Rijke et al. 2012) and entrepreneurs are often described as having leadership as a key quality (Christopoulos 2006; Weible 2007). Several examples in water governance literature also highlight ways in which leaders engage in behavior characteristic of policy entrepreneurs. For example, Bettini et al. (2015) suggest that leadership is essential for framing issues, building relationships, exploiting opportunities, garnering resources, and maintaining momentum for water governance systems to have the adaptive capacity to maintain resilient socio-ecological systems. Similarly, Farrelly and Brown (2011) recognize leadership as necessary for transition to sustainable urban water practices because it promotes organizational experimentation and learning.

Discovering Important Policy Actors through Network Analysis

Measures of network position are often taken "off the shelf" to operationalize and identify roles that are key to policy outcomes. Researchers have used network analytic techniques, for example, to study how important actors carry out operational goals, such as enabling resource exchange (Crona and Bodin 2006), reducing conflict (Varone et al. 2019), and promoting learning (Hysing and Olsson 2008). The key challenge of network operationalization is to find measures that adequately reflect the defining characteristics of important policy actors.

Researchers most commonly rely on two positional measures to operationalize importance: *degree centrality* and *betweenness centrality* (Freeman 1979). Degree centrality refers to the number of ties an actor (i.e., *ego*) shares with adjacent others (i.e., *alters*) in the network (Knoke and Yang 2008). In network terms, entrepreneurs may be identified through their degree centrality. The ties they form can be instrumental for a variety of reasons, such as coordinating action through the sharing of diverse information and resources, or by mobilizing more stakeholders to take action needed for influencing the policy agenda (Sabatier and Jenkins-Smith 1999). While the water governance literature tends not to describe policy entrepreneurs as responsible for transitions in water management technologies, "leaders" frequently appear in case studies as change agents that work with others to achieve policy objectives (Farrelly and Brown 2011; Bettini et al. 2015). As such, degree centrality may be, similarly, appropriate for operationalizing this role in the policy network.

Some scholars have used degree centrality to characterize important actors and how they influence the policy process. Ingold (2011), for example, combines SNA with multi-criteria analysis and finds that actors with high betweenness centrality have moderate to centrist beliefs between pro-economy and pro-ecology positions, and are well suited to enable compromise in a hurting stalemate between competing advocacy coalitions. Another study finds that policy actors with high degree centrality have comparatively higher success in securing passage of anti-fracking policies when compared to those that are less central in the network (Arnold et al. 2017).

Betweenness centrality refers to the frequency with which a given actor stands between the shortest path between pairs of actors in the full network (Knoke and Yang 2008). Policy brokers can be measured in social networks by their betweenness centrality, following the assumption that they are linking sparsely or completely disconnected subgroups. These structural positions enable strategic leverage over other network participants by controlling flows of non-redundant information and resources (Burt 2009).

Ingold and Varone (2012), for example, measure brokerage through reports of reputational power, or "importance" in the Swiss climate policy subsystem, but find that not all respondents share the same opinion about the importance of policy actors that have the highest betweenness centrality in the network. Another study by Christopoulos and Quaglia (2009) uses surveys, interviews, and graph theoretic approaches to identify most influential actors and their means of influence in the EU banking policy domain.

Challenges of Network Measurement

Despite the utility of network analysis, empirical findings show a need to address incongruence between relational, network measures of centrality and qualitative reports of importance. For example, Christopoulos and Quaglia (2009) find that many of the actors reported as important—which is operationalized as prominence in lobbying activities relating to banking legislation—tend to not have *honest* brokerage roles in the network.[1] Christopoulos and Ingold (2015) also draw insights about measurement limitations in both the 2009 case and the analysis by Ingold and Varone (2012). They find that despite using additional measures of centrality—including Bonacich power,[2] Burt's effective size and constraint,[3] and honest brokerage—there is no clear correspondence with descriptive importance or impact on the policy process.

One challenge to measurement can stem from data collection, especially if important roles in policy settings are too broadly or narrowly specified. For example, asking respondents to nominate those of "importance" in the policy process can elicit a sample of actors that occupy a wide variety of prominent roles. Yet, this may not include those

---

[1] Honest brokerage is defined as a relational position in the network structure, where the actor of interest serves as the only intermediary between other actors (Borgatti et al. 2002).

[2] The concept of Bonacich Power proposes that the amount of power one has is a function of the power of those to whom she is connected (Bonacich 1987).

[3] Burt's (1992) effective size refers to the network size minus redundancy in the network, and constraint refers to an index that measures the redundancy of one's contacts.

participating as brokers or entrepreneurs. Alternately, narrowly defining what activity signals behavior of a broker or entrepreneur can make multiplexity of actions inherent to each role go undetected (Christopoulos and Ingold 2011).

Respondent error can also impact the accuracy of measurement. Some individuals may only nominate those with whom they engage frequently (Bernard et al. 1984) and fail to list others that indeed occupy important roles in the network. Empirical research on political decision making has shown, for example, that individuals in policy networks may attribute more power to those with whom they closely collaborate during the decision-making processes, regardless of that collaborator's formal authority, intensity of participation, or centrality in the network (Fischer and Sciarini 2015). In other cases, individuals may have differing *cognitive social structures*—some may have limited knowledge of extant relationships, whereas others are comparatively well informed (Krackhardt 1987).

Thus, evaluating the extent to which network measures reflect important policy actors requires a means of validation. Measurement is never precise in social sciences (King et al. 1994). So, while network measures may be valid in theoretical terms, comparing these against expert knowledge can improve reliability.

Data: Networks in Urban Water Governance in Arizona, USA

This case focuses on policy stakeholders participating in urban water governance in Tucson, Arizona, USA. Traditionally, the city has relied largely on the use of groundwater, which it supplies to customers through a centralized, piped system (Tucson Water 2000). Tucson is located in a semi-arid region that receives little rainfall and, like most of the state, has a history of groundwater overdraft (Megdal 2012). In 1980, the state legislature passed the Groundwater Management Act to establish a series of management plans that would guide conservation areas (called "Active Management Areas") in achieving a *safe yield*[4] by 2025 (ADWR 2002).

---

[4] This term refers to annual groundwater withdrawals that do not exceed the amount recharged (ADWR 2016).

The city began to adopt additional measures to reduce the risk of over-extraction. Within the past three decades, Tucson has expanded its portfolio to include innovative approaches, such as rainwater harvesting, graywater applications, and the use of water-efficient appliances (City of Tucson and Pima County 2009; Tucson Water 2013; Davis 2014; Hester et al. 2015).

Both the Tucson and surrounding Pima County governments have been highly involved in the adoption and implementation of Tucson's water policies. More recently, however, diversification in water management approaches has brought new stakeholders into the policy domain. A growing group of non-profit and private organizations, for example, have become involved in spearheading and supporting innovative programs and practices in the local water governance system. We expect, then, narrative reports and network measures to detect these actors in this policy setting.

Data Collection

To gather data for our analysis, we used a systematic Google search of gray literature published online.[5] Gray literature refers to resources such as news articles, white papers, and other media that are often produced outside commercial venues, not standardly distributed, and not found in standard bibliographic retrieval systems (Tillett and Newbold 2006; Mahood et al. 2014). Analysis of content, such as gray literature, has long been an efficient and low-cost tool for researchers (Laumann et al. 1987) and is becoming increasingly common in different policy studies (Crow 2010; Pierce 2011; Hayes and Scott 2017).

We identified key water-management areas by reviewing expenditures outlined in the US Bureau of the Census *Government Finance and Employment Classification Manual* (US Census Bureau 2006). Following the guidance of water experts in hydrology, planning, and natural resources, we then organized the activities into seven terms representing

---

[5] Example of media found in the gray literature search included ordinances, news articles, informational pamphlets, presentations, user guides or manuals, white papers, technical reports, archives, and meeting minutes.

functional domains of water management.[6] To identify gray literature pertaining to these domains in the Tucson water governance system, we paired each term with "Tucson, Arizona." The prevalence of recently published materials in our results was addressed by conducting the search with each combination for each year, from 2007 to 2017. Given limited time and human capital, we only selected the first ten relevant results for each combination and year.[7] Some of the resources included links to other relevant content; this created an opportunity for informal snowball sampling. In total, the corpus included 844 gray literature documents. From the content of this corpus, we identified 784 water-related policies and 440 actors.

The Tucson Water Governance Network

From the gray literature, we derived a network of organizations[8] that coordinated on water management strategies in Tucson in 2017.[9] Network actors included those of whom the gray literature described as implementing or promoting water-related policies or programs. We operationalized *coordination* using an event-based approach, where two actors participating in the same event,[10] were assumed to have this type of relation (Laumann et al. 1989). The event-based approach was appropriate for delineating the boundaries of the network because it was impossible

---

[6] Functional domains included *stormwater, water supply, groundwater, flood, wastewater, water quality*, and *water conservation*.

[7] By "relevant" we refer to whether the gray literature resource actually entailed information pertaining to water management in or around Tucson. Other documents that were excluded from the first ten results were repeats from earlier searches; each repeat was substituted with the next available, relevant search result.

[8] Organizations included public, non-profit, private sectors, as well as research organizations, media, and interest groups. This was with the exception of one individual without organizational affiliation, who was also included.

[9] Although we had data for 2007–2017, there was insufficient information in gray literature narratives on when actors assumed or left brokerage and entrepreneurship roles.

[10] By events, we refer to programs and policies. For example, in the Tucson water governance system, this includes the local water utility's Rainwater Harvesting Rebate Program, designation demonstration sites throughout the city, and the City's Green Streets Active Practice Guidelines that require development of GI and LID features in all road construction or updates (City of Tucson 2009; Tucson Department of Transportation 2013).

to know relevant policy actors a priori (Laumann and Knoke 1987), and because much of the gray literature content did not offer detailed information on the nature of relationships between actors of interest.

Measurement of Actor Importance

Collectively, these ties comprised a unipartite (actor-actor) network of 268 actors that coordinated on at least one of 526 events in 2017. The unipartite network allows us to observe centrality and brokerage among coordinating actors (i.e., stakeholder organizations and some individuals without affiliate organizations), whereas a bipartite (actor-event) network would convey information about centrality as the degree of an actor's involvement in events, or participation in a set of events that may each have a group of unique participants. In this case, the presence of entrepreneurship and brokerage among coordinating actors was best conceptually represented through the unipartite structure. From this network, we derived the first of two independent importance measures: degree centrality and betweenness of organizations and individuals in the water policy network (Wasserman and Faust 1994).

The second measure came from coding media accounts of organizations as brokers or entrepreneurs. This was done by recording explicit mentions of actors as brokers or entrepreneurs, as well as mentions of organizational behavior that is consistent with these roles. This included self-reports, such as organization websites that reviewed their respective accomplishments, as well as third-person reports—as generated through sources such as media sites—on the policy behavior in the water governance network. We conceptualized entrepreneurial behavior as actions to promote change, such as advocating for and participating in activities to support the adoption or implementation of programs or innovative technologies practiced therein. Brokerage behavior was broadly construed, entailing the occupation of some role between two other organizations or events that were also constituent to the urban water governance network. For example, we conceptualized a local non-profit—the Watershed Management Group (WMG)—as a broker because it was reported as acting on behalf of the City Council to work with the surrounding

County's flood control organization to address localized flooding challenges within Tucson (Watershed Management Group & Pima County Regional Flood Control District 2015).

Data Analysis

We use a point-biserial correlation—a variation on the Pearson's r—to compare reports of importance in the policy system to network measurements of relational importance. This technique is appropriate for estimating the correlation coefficient because the network measures of centrality are continuous, but the nominations of brokers and entrepreneurs are dichotomous (each role is correlated with the associated network measure). Like Pearson's r, a point biserial r estimates the correlation coefficient between two variables, X and Y, where 1 indicates a completely positive, linear correlation, 0 indicates no correlation, and -1 signifies a totally negative linear correlation (Lee Rodgers and Nicewander 1988; Pearson and Francis 1895; Kurtz and Mayo 1979).[11]

Approximately 8% of the network comprised peripheral actors that each had one tie to more central network members in the structure. As such, it was necessary to take the natural log transformation of the data. Figure 11.1 shows that log-transforming the degree centrality data satisfied the assumption of a normal distribution, save a few outlier observations at the lower end of the theoretical quantile.

We also log-transformed degree centrality data for actors nominated in the gray literature as entrepreneurs, and then also for actors *not* described as fulfilling these roles. In the entrepreneur and non-entrepreneur conditions, there were several outliers with a sample quantile of 0. This likely indicates many peripheral actors that originally had a degree centrality score of 1 received a score of 0 after the natural log transformation.

To assess whether there was equality of variance for the different entrepreneur conditions, we then used a Levene's test, which indicates the

---

[11] Assumptions of the point-biserial correlation require that (1) the continuous variable X is normally distributed, and has a normal distribution for each level of the dichotomous variable, and (2) that the continuous variable X has equal variances for each level of the dichotomous variable (Pett 2015).

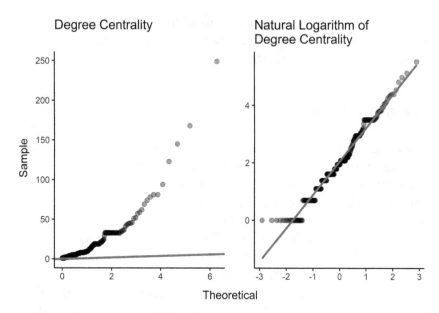

Fig. 11.1 Q–Q Plots of degree centrality and log transformations of degree centrality for the whole network

homogeneity of variance. The p value is greater than $\alpha = 0.05$ and is insignificant. This means that the data satisfy the second assumption of the point-biserial correlation, that X has equal variances for both the entrepreneur and non-entrepreneur conditions.

Unlike degree centrality, betweenness centrality measures failed to satisfy the assumption of normal distribution. This was because more than 70% of the network members had a betweenness centrality score of 0, suggesting that they occupied peripheral positions in the network. The skew in the distribution was extensive enough that transforming the data did little to improve the distribution. However, the observations with betweenness centrality greater than or equal to 1 were normally distributed (Fig. 11.2). A Breusch-Pagan test also showed that these observations were homoscedastic. Thus, we decided to subsample these observations and examine their relationship with gray literature nominations and reports of brokerage and reports of behavior that satisfies a brokerage role.

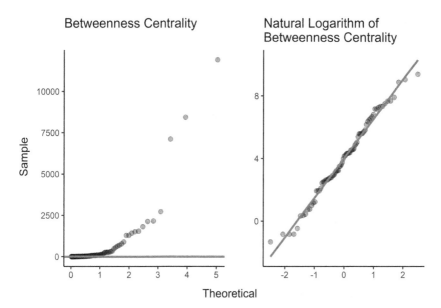

Fig. 11.2 Q–Q plot of betweenness centrality (left) and the distribution of log-transformed observations with betweenness centrality greater than or equal to 1

Table 11.1 Correlation between network centrality and reports of important actors

	Reported importance	
Network measures of importance	Broker (subsample)	Entrepreneur
Degree centrality	0.106[†]	0.280***
Betweenness centrality	0.256*	0.329**

The subsample of betweenness centrality included only observations with scores greater than or equal to 1
[†]$p \leq 0.10$; *$p \leq 0.05$; **$p \leq 0.01$; ***$p \leq 0.001$

There is a significant and positive correlation between entrepreneurship roles and both degree and betweenness centrality. Also, there is evidence that the brokerage-betweenness centrality correspondence is slightly stronger than the correspondence between brokerage and degree centrality (Table 11.1).

Discussion

This case shows how point-biserial correlation is an effective approach for assessing the validity of measurement in water governance. Also, the strength of association between diverse measures of importance facilitates a critical assessment of operationalization and considerations for future work. From these findings we draw three key insights. First, unique circumstances in the policy setting—such as levels of conflict or the presence of highly involved stakeholders—can impact both network measures and the extent of gray literature reporting on specific policy roles. Second, bias in reporting and the types of content from which data are derived might impact measurement precision. Third, event-based analysis is efficient but offers lower precision in detection of policy brokers than other data-collection techniques. After discussing each of these points at greater length, we critically assess the utility of correlating descriptive network statistics with alternative measures as a validation strategy. Lastly, we present concluding thoughts.

Circumstances in the Policy Setting

The point biserial correlation directed our attention to how measurement and interpretation of importance can be linked to context-specific factors. Key themes we observed in the Tucson water governance case were potential lack of policy conflict, prominence of government organizations, and—albeit to a lesser extent—high activity among a small number of actors from the non-profit and academic sectors.

We observed in the Tucson water governance case, for example, that the strength of correlation between reported brokerage and network measures of betweenness centrality is small (0.256). This could suggest that, while some actors participated in more than one policy or program, this brokerage between groups of largely disconnected network members does not necessarily signal the presence of policy conflict. Of the 268 actors identified, only 5% were reported having any kind of adversarial relationship with others in the network. The network structure also showed low levels of modularity (0.455) indicating the presence of a few densely connected groups of actors that were isolated or that shared few ties with other groups (Newman 2006). This primarily occurred at the core of the network, with many peripheral actors that had low degree-centrality scores. In this type of structure, brokers can transfer novel information to peripheral actors that seek to develop new technologies and advance sustainability (Henry and Vollan 2014).

Another trend in the gray literature showed that most policy actors coded as entrepreneurs and/or brokers were from the government sector. These government actors also demonstrated long-term involvement through spearheading programs in Tucson's water governance. Tucson Water, for example, introduced its well-known "Beat-the-Peak" educational program in 1987 to encourage the city's water consumers to reduce summer peak demand (City of Tucson and Pima County 2009); the Pima County Regional Wastewater Reclamation District began its Biosolid Land Application Program in 1980, which continues to provide biosolids recycled from wastewater as a source of fertilizer and soil conditioner in the agricultural industry (Pima County Wastewater Reclamation 2016); the Pima Association of Governments adopted its Area Water Quality Management Plan in 1978 to fulfill requirements of the Clean Water Act (Pima Association of Governments 2012), identified problem areas for point and non-point pollution, and recommended solutions and alternatives to address these challenges; and the Tucson Department of Transportation introduced its 1988 Interim Watercourse Improvement Policy to preserve vegetation adjacent to urban waterways (Cleveland 2013).

The University of Arizona and the WMG (discussed earlier) also showed top centrality scores and were reported as important actors. Like the important government stakeholders, the gray literature indicated that the University had also been active for several decades. One of its earlier water management strategies was its 1985 development of Casa del Agua and the Arizona Desert House, which served as demonstration sites to educate the public about water conservation techniques, such as use of graywater, efficient appliances, rainwater harvesting, and drought-tolerant plants (Sheikh 2009). Although the Watershed Management Group was not founded until 2003, it continued to play a key leadership role in the adoption and implementation of green infrastructure policies through 2017. The gray literature reported the organization's responsibility for 13 different water conservation programs, events, and other approaches. Some of these included its rainwater harvesting demonstration site, workshops, and certification program (City of Tucson 2009, 2017; Watershed Management Group n.d.).

These public, non-profit, and academic actors that had the highest centrality scores among those nominated as important were also very well connected in the network. In 2017, each of the most central organizations participated in a relatively high percentage of that year's programs and events (5%) as compared to less-central stakeholders in the Tucson water policy network (0.7%). Of the important actors, Tucson Water and the Pima County Regional Wastewater and Reclamation District were most involved in the network, participating in 23% and 16% of all water programs and policies, respectively. This was of little surprise, given that these water service providers are responsible for supply and wastewater management in the Tucson metropolitan area. The University of Arizona and the Tucson Department of Transportation both participated in comparatively fewer events (each 7%), and the WMG and the Pima Association of Governments each participated in 5%.

Other stakeholders that were nominated as important, but that did not show high centrality, were often connected to innovative policies. Much like the Watershed Management Group, several non-profit organizations that were nominated as leaders were also responsible for introducing green infrastructure approaches, such as the Neighborhood Scale Stormwater Harvesting Program, introduced by Tucson Clean and Beautiful (Tucson Water n.d.), the Sonoran Environmental Research Institute's loan program for low-income residents to install passive or active rainwater harvesting systems (Davis 2016), the Tucson Botanical Garden's green stormwater infrastructure workshops and installment (Phillips 2005; Cleveland 2013), and the Santa Cruz River Initiative, an effort spearheaded by the Sonoran Institute to improve watershed conditions (WRRC 2008).

Finally, the gray literature also directed much attention to Brad Lancaster, a leader in piloting several water conservation efforts, organizing local stakeholders, and advocating for policy innovation. Earlier efforts included his 1996 development of the Annual Dunbar Spring Neighborhood Rain, Tree, and Carbon Planting event, which continues to bring residents together to install and maintain street-side passive rainwater harvesting features (Permaculture Research Institute 2016). Lancaster is also well known historically for making curb cuts prior to city approval to facilitate better infiltration of street-side stormwater; the City of Tucson later institutionalized this approach in its Curb Cut Standards (Riley 2013).

Bias in Reporting and Content Type

Figure 11.3 shows that (as would be expected) several actors reported as entrepreneurs in the gray literature tend to have the greatest degree centrality among the different network actor roles. Yet, there are also many reported entrepreneurs with degree centrality no greater than those reported to be brokers, those occupying both brokerage and entrepreneurial roles, or those that the gray literature did not report.

The patterns we observe may be explained by biased reporting and, relatedly, *type* of gray literature content. The wide range of degree centrality scores for the reported entrepreneurs in Fig. 11.3 may indicate that

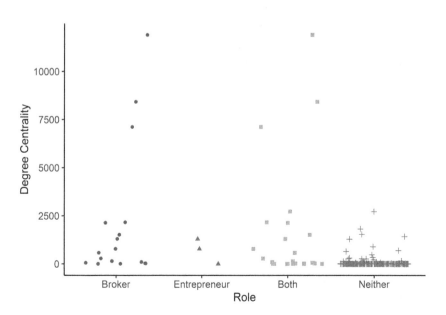

Fig. 11.3 Frequency plot of actor type and degree centrality

media is directing additional attention to a few organizations with well-established reputations in the policy domain, or that are involved in the latest policies or problems (Yi and Scholz 2016). This may also reflect the prevalence of government reports in the corpus. These types of documents often detailed the latest policy innovations and information about actors responsible for implementation of current programs and events. Alternately, a much smaller portion of the documents discussed policy actors advocating for adoption of a policy, or the processes leading up to a policy innovation.

Limits of Event-Based Analysis

Event-based analysis for delimiting network boundaries is useful when there is no prior knowledge about what set of actors exists. While scholars have used this approach for several decades (Dahl 2005), it nonetheless presents key limitations. One issue is properly identifying which actors have relations. Assuming coordination through joint event participation can lead to overestimation of network density. Also, higher density can artificially inflate the number of shortest paths, decreasing the betweenness centrality of some organizations that might otherwise be identified as holding key brokerage positions in the network. Related to this, identifying coordination ties through a limited set of events could also exclude relevant actors from the network (Laumann et al. 1989). For example, analysis of events outside of water programs and policies—such as social gatherings or meetings—might detect other important exchanges that are useful for understanding dynamics in the policy network. This may suggest that researchers need to use an additional "relational" strategy, with closer attention to specifying key interactions (Marsden 2011), such as joint advocacy, resource sharing, or information exchange.

Conclusion

The point biserial correlation illustrates a promising tool for validity assessment because it can highlight data features that should receive further evaluation. In this case, for example, variation in the strength of the linear association between the network measurements and gray literature reports highlighted outstanding data-related questions, such as (1) what operationalization of gray literature reports offers adequate precision? and (2) under what circumstances is event-based analysis sufficient for network construction? These areas of inquiry highlight potential avenues for future research, such as using discrete, ternary, and quaternary data in place of binary measurements to correlate with network measures. Operationalizing ties through rule-based text analysis could also minimize error in tie detection.

In water governance, extracting valid measurements of brokerage and policy entrepreneurship from quantitative network centrality scores lends insights into how to manage key governance challenges. For instance, brokers and entrepreneurs may be in a particularly advantageous position to champion new technical innovations that promote resilience. The position of brokers may also highlight fragmentations in the network where policy participants with complementary resources and skillsets would benefit from collaboration. It could also illustrate how systems are adapting and promoting resilience to climatic pressures for water governance, including long-term drought, flooding, and high variability in precipitation.

This validation approach is also well suited for application across diverse policy contexts, enabling comparative analyses. Estimating the correlation of mentions of organizational importance across different locations, for example, could illustrate variation in local reporting or availability of content. It might also speak to unique socio-political factors that are critical in shaping the policy process.

Finally, the correlation of network measures with different forms of descriptive statistics can be useful for assessing the validity of other constructs and structural features. One promising direction is further examination of measuring social capital (Lin 2017; Berardo and Scholz 2010; Henry et al. 2011). Namely, correlating reported theoretical variables with different dimensions of multiplex network ties can help unpack the social capital construct.

This study allows us to draw several lessons from testing the validity of theoretical and network-based measurements of policy actors. To recapitulate findings, the point biserial correlation shows a significant positive relationship between entrepreneurs and degree centrality, as well as betweenness centrality. There is also a significant, positive correlation between brokers and degree centrality and betweenness. The results are consistent with the theory suggesting that actor importance is reflected in, among other things, network position. This research illustrates the ongoing need to carefully compare network centrality measures against theoretical concepts for which network centrality is often used as a proxy. The coding of gray literature is one way to accomplish this goal. This approach demonstrates an ability to reveal context-specific factors that warrant more in-depth analysis, such as bias in media reporting, differential institutional structures, or unique socio-political dynamics.

References

ADWR. (2002). *Securing Arizona's Water Future: Overview of the Arizona Groundwater Management Code.*

ADWR. (2016). Active Management Areas (AMAs) & Irrigation Non-Expansion Areas (INAs). https://new.azwater.gov/ama

Arnold, G., Long, L. A. N., & Gottlieb, M. (2017). Social Networks and Policy Entrepreneurship: How Relationships Shape Municipal Decision Making about High-Volume Hydraulic Fracturing. *Policy Studies Journal: The Journal of the Policy Studies Organization, 45*(3), 414–441.

Berardo, R., & Scholz, J. T. (2010). Self-Organizing Policy Networks: Risk, Partner Selection, and Cooperation in Estuaries. *American Journal of Political Science, 54*(3), 632–649.

Bernard, H. R., Killworth, P., Kronenfeld, D., & Sailer, L. (1984). The Problem of Informant Accuracy: The Validity of Retrospective Data. *Annual Review of Anthropology, 13*(1), 495–517.

Bettini, Y., Brown, P. R., & de Haan, F. J. (2015). Exploring Institutional Adaptive Capacity in Practice: Examining Water Governance Adaptation in Australia. *Ecology and Society, 20*(1), 47.

Bonacich, P. (1987). Power and Centrality: A Family of Measures. *The American Journal of Sociology, 92*(5), 1170–1182.

Borgatti, S. P., Everett, M. G., & Freeman, L. C. (2002). *Ucinet for Windows: Software for Social Network Analysis.* Harvard, MA: Analytic Technologies.

Brouwer, S., & Biermann, F. (2011). Towards Adaptive Management: Examining the Strategies of Policy Entrepreneurs in Dutch Water Management. *Ecology and Society, 16*(4), 5.

Brown, R. R. (2008). Local Institutional Development and Organizational Change for Advancing Sustainable Urban Water Futures. *Environmental Management, 41*(2), 221–233.

Burt, R. (1992). *The Social Structure of Competition*. Cambridge, MA: Harvard University Press.

Burt, R. S. (2009). *Structural Holes: The Social Structure of Competition*. Cambridge, MA: Harvard University Press.

Christopoulos, D. (2006). Relational Attributes of Political Entrepreneurs: A Network Perspective. *Journal of European Public Policy, 13*(5), 757–778.

Christopoulos, D., & Ingold, K. (2011). Distinguishing Between Political Brokerage & Political Entrepreneurship. *Procedia—Social and Behavioral Sciences, 10*, 36–42.

Christopoulos, D., & Ingold, K. (2015). Exceptional or Just Well Connected? Political Entrepreneurs and Brokers in Policy Making. *European Political Science Review, 7*(3), 475–498.

Christopoulos, D., & Quaglia, L. (2009). Network Constraints in EU Banking Regulation: The Capital Requirements Directive. *Journal of Public Policy, 29*(2), 179–200.

City of Tucson. (2009). *Rainwater Harvesting Rebate Program*. Retrieved from https://www.tucsonaz.gov/water/rwh-rebate.

City of Tucson. (2017). *Water Matters*.

City of Tucson, and Pima County. (2009). *City of Tucson and Pima County Water Conservation Technical Paper*.

Cleveland, J. (2013). *Regional Green Stormwater Infrastructure Survey*. Progress Report.

Crona, B., & Bodin, Ö. (2006). What You Know Is Who You Know? Communication Patterns Among Resource Users as a Prerequisite for Co-Management. *Ecology and Society, 11*(2), 7.

Crow, D. A. (2010). Policy Entrepreneurs, Issue Experts, and Water Rights Policy Change in Colorado. *The Review of Policy Research, 27*(3), 299–315.

Dahl, R. A. (2005). *Who Governs?: Democracy and Power in an American City. Yale Studies in Political Science* (2nd ed.). New Haven, CT: Yale University Press.

Davis, T. (2014). City Council Expands Water-Harvesting Rebates. *Arizona Daily Star.*

Davis, T. (2016, July). Poor People Left out of Tucson Water Harvesting Rebates. *Arizona Daily Star*. Retrieved from http://tucson.com/news/local/low-income-residents-left-out-of-tucson-water-harvesting-rebates/article_acb89804-e9b3-5929-adf3-ea9d443ab089.html.

Ernstson, H., van der Leeuw, S. E., Redman, C. L., Meffert, D. J., Davis, G., Alfsen, C., & Elmqvist, T. (2010). Urban Transitions: On Urban Resilience and Human-Dominated Ecosystems. *Ambio, 39*(8), 531–545.

Farrelly, M., & Brown, R. (2011). Rethinking Urban Water Management: Experimentation as a Way Forward? *Global Environmental Change: Human and Policy Dimensions, 21*(2), 721–732.

Feiock, R. C. (2013). The Institutional Collective Action Framework. *Policy Studies Journal: The Journal of the Policy Studies Organization, 41*(3), 397–425.

Fischer, M., & Sciarini, P. (2015). Drivers of Collaboration in Political Decision Making: A Cross-Sector Perspective. *The Journal of Politics, 78*(1), 63–74.

Freeman, L. C. (1979). Centrality in Social Networks Conceptual Clarification. *Social Networks, 1*(3), 215–239.

Gunderson, L., & Light, S. S. (2006). Adaptive Management and Adaptive Governance in the Everglades Ecosystem. *Policy Sciences, 39*(4), 323–334.

Hayes, A. L., & Scott, T. A. (2017). Multiplex Network Analysis for Complex Governance Systems Using Surveys and Online Behavior. *Policy Studies Journal: The Journal of the Policy Studies Organization, 46*(2), 327–353.

Henry, A. D., & Vollan, B. (2014). Networks and the Challenge of Sustainable Development. *Annual Review of Environment and Resources, 39*(1), 583–610.

Henry, A. D., Lubell, M., & McCoy, M. (2011). Belief Systems and Social Capital as Drivers of Policy Network Structure: The Case of California Regional Planning. *Journal of Public Administration Research and Theory, 21*(3), 419–444.

Hester, B., Ogata, I., Wittwer, G., Meir, M., Pope, J., Zucker, C., & Light, M. (2015). Low Impact Development and Green Infrastructure Guidance Manual. https://doi.org/10.1017/CBO9781107415324.004.

Hysing, E., & Olsson, J. (2008). Contextualising the Advocacy Coalition Framework: Theorising Change in Swedish Forest Policy. *Environmental Politics, 17*(5), 730–748.

Ingold, K. (2011). Network Structures within Policy Processes: Coalitions, Power, and Brokerage in Swiss Climate Policy. *Policy Studies Journal: The Journal of the Policy Studies Organization, 39*(3), 435–459.

Ingold, K., & Varone, F. (2012). Treating Policy Brokers Seriously: Evidence from the Climate Policy. *Journal of Public Administration Research and Theory*, *22*(2), 319–346.

Karvonen, A. (2010). Metronatural™: Inventing and Reworking Urban Nature in Seattle. *Progress in Planning*, *74*(4), 153–202.

Keremane, G. (2015). Role of Sustainability Policy Entrepreneurs in Building Water-Sensitive Cities to Respond to Climate Change: A Case Study in Adelaide, Australia. In S. Shrestha, A. K. Anal, P. A. Salam, & M. van der Valk (Eds.), *Managing Water Resources under Climate Uncertainty: Examples from Asia, Europe, Latin America, and Australia* (pp. 359–375). Cham: Springer International Publishing.

King, G., Keohane, R. O., & Verba, S. (1994). *Designing Social Inquiry: Scientific Inference in Qualitative Research*. Princeton, NJ: Princeton University Press.

Kingdon, J. W. (1995). *Agendas, Alternatives, and Public Policies* (2nd ed.). New York, NY: Longman.

Knoke, D., & Yang, S. (2008). *Social Network Analysis*. Thousand Oaks, CA: Sage.

Krackhardt, D. (1987). Cognitive Social Structures. *Social Networks*, *9*, 109–134.

Kurtz, A. K., & Mayo, S. T. (1979). *Statistical Methods in Education and Psychology*. New York, NY: Springer New York.

Laumann, E. O., & Knoke, D. (1987). *The Organizational State: Social Choice and National Policy Domains*. University of Wisconsin Press.

Laumann, E. O., Knoke, D., & Kim, Y. (1987). Event Outcomes. *The Organizational State. Social Choice in National Policy Domains*, 271–287.

Laumann, E. O., Marsden, P. V., & Prensky, D. (1989). The Boundary Specification Problem in Network Analysis. In L. C. Freeman, D. R. White, & A. K. Ronney (Eds.), *Research Methods in Social Network Analysis* (pp. 61–87). Fairfax, VA: George Mason University Press.

Lee Rodgers, J., & Nicewander, W. A. (1988). Thirteen Ways to Look at the Correlation Coefficient. *The American Statistician*, *42*(1), 59–66.

Lin, N. (2017). Building a Network Theory of Social Capital. In R. Dubos (Ed.), *Social Capital* (pp. 3–28). New York, NY: Routledge.

Mahood, Q., Van Eerd, D., & Irvin, E. (2014). Searching for Grey Literature for Systematic Reviews: Challenges and Benefits. *Research Synthesis Methods*, *5*(3), 221–234.

Marsden, P. V. (2011). Survey Methods for Network Data. In J. Scott & P. J. Carrington (Eds.), *The SAGE Handbook of Social Network Analysis* (pp. 370–388). London: Sage Publications.

Megdal, S. B. (2012). The Role of the Public and Private Sectors in Water Provision in Arizona, USA. *Water International, 37*(2), 156–168.

Mintrom, M., & Norman, P. (2009). Policy Entrepreneurship and Policy Change. *Policy Studies Journal: The Journal of the Policy Studies Organization, 37*(4), 649–667.

Mintrom, M., & Vergari, S. (1998). Policy Networks and Innovation Diffusion: The Case of State Education Reforms. *The Journal of Politics, 60*(1), 126–148.

Newman, M. E. J. (2006). Modularity and Community Structure in Networks. *Proceedings of the National Academy of Sciences of the United States of America, 103*(23), 8577–8582.

Olsson, P., Gunderson, L. H., Carpenter, S. R., Ryan, P., Lebel, L., Folke, C., & Holling, C. S. (2006). Shooting the Rapids: Navigating Transitions to Adaptive Governance of Social-Ecological Systems. *Ecology and Society, 11*(1), 18.

Olsson, P., Folke, C., Galaz, V., Hahn, T., & Schultz, L. (2007). Enhancing the Fit through Adaptive Co-Management: Creating and Maintaining Bridging Functions for Matching Scales in the Kristianstads Vattenrike Biosphere Reserve, Sweden. *Ecology and Society, 12*(1), 28.

Pearson, K., & Francis, G. (1895). VII. Note on Regression and Inheritance in the Case of Two Parents. *Proceedings of the Royal Society of London, 58*(347–352), 240–242.

Permaculture Research Institute. (2016). *Dryland Harvesting Home Hacks Sun, Rain, Food & Surroundings*. Retrieved from https://permaculturenews.org/2016/08/25/dryland-harvesting-home-hacks-sun-rain-food-surroundings/.

Pett, M. A. (2015). *Nonparametric Statistics for Health Care Research: Statistics for Small Samples and Unusual Distributions* (2nd ed.). Thousand Oaks, CA: Sage.

Phillips, A. A. (2005). *Water Harvesting—Guidance Manual*. City of Tucson, Arizona. Retrieved from https://www.tucsonaz.gov/files/transportation/2006 WaterHarvesting.pdf.

Pierce, J. J. (2011). Coalition Stability and Belief Change: Advocacy Coalitions in US Foreign Policy and the Creation of Israel, 1922–44. *Policy Studies Journal: The Journal of the Policy Studies Organization, 39*(3), 411–434.

Pima Association of Governments. (2012). *Top 10 Environmental Issues*. PAG Environmental Planning Advisory Committee (EPAC). Retrieved from https://www.pagnet.org/documents/top10envissues2012.pdf.

Pima County Wastewater Reclamation. (2016). *2016 Wastewater Facility Plan*. Tucson, Arizona. Retrieved from https://webcms.pima.gov/UserFiles/

Servers/Server_6/File/Government/Wastewater%20Reclamation/Publibations/FacilityPlan_2016.pdf.

Rijke, J., Brown, R., Zevenbergen, C., Ashley, R., Farrelly, M., Morison, P., & van Herk, S. (2012). Fit-for-Purpose Governance: A Framework to Make Adaptive Governance Operational. *Environmental Science and Policy, 22*, 73–84.

Riley, M. (2013). Locals Promote Rainwater Harvesting In Creative Forms. *Arizona Public Media*. Retrieved from https://www.azpm.org/s/15996-harvesting-the-rain/.

Rouillard, J. J., Heal, K. V., Ball, I., & Reeves, A. D. (2013). Policy Integration for Adaptive Water Governance: Learning from Scotland's Experience. *Environmental Science and Policy, 33*, 378–387.

Sabatier, P. A., & Jenkins-Smith, H. C. (1993). *Policy Change and Learning*. Boulder, CO: Westview Press.

Sabatier, P. A., & Jenkins-Smith, H. C. (1999). The Advocacy Coalition Framework: An Assessment. In P. A. Sabatier (Ed.), *Theories of the Policy Process* (pp. 117–166). Boulder, CO: Westview Press.

Sharma, A. K., Cook, S., Tjandraatmadja, G., & Gregory, A. (2012). Impediments and Constraints in the Uptake of Water Sensitive Urban Design Measures in Greenfield and Infill Developments. *Water Science and Technology: A Journal of the International Association on Water Pollution Research, 65*(2), 340–352.

Sheikh, B. (2009). *White Paper on Graywater*. Retrieved from http://www.azwater.gov/azdwr/WaterManagement/documents/GraywaterWhitePaperFinal.pdf.

Sitas, N., Reyers, B., Cundill, G., Prozesky, H. E., Nel, J. L., & Esler, K. J. (2016). Fostering Collaboration for Knowledge and Action in Disaster Management in South Africa. *Current Opinion in Environmental Sustainability, 19*, 94–102.

Tillett, S., & Newbold, E. (2006). Grey Literature at The British Library: Revealing a Hidden Resource. *Interlending and Document Supply, 34*(2), 70–73.

Tompkins, E., Adger, W. N., & Brown, K. (2002). Institutional Networks for Inclusive Coastal Management in Trinidad and Tobago. *Environment & Planning A, 34*(6), 1095–1111.

Tucson Department of Transportation. (2013). *Engineering Division Active Practice Guidelines*. Retrieved from https://watershedmg.org/sites/default/files/documents/city-of-tucson-green-streets-active-practice-guidelines.pdf.

Tucson Water. (2000). *Water Plan: 2000–2050*.

Tucson Water. (2013). *2012 Update: Water Plan 2000–2050*. Retrieved from http://www.tucsonaz.gov/files/water/docs/2012_Update_Water_Plan_2000-2050.pdf.

Tucson Water. (n.d). *Neighborhood Scale Stormwater Harvesting*.

US Census Bureau. (2006). *Government Finance and Employment Classification Manual*.

Varone, F., Ingold, K., & Fischer, M. (2019). Policy Networks and the Roles of Public Administrations. In A. Ladner, N. Soguel, Y. Emery, S. Weerts, & S. Nahrath (Eds.), *Swiss Public Administration: Making the State Work Successfully* (pp. 339–353). Cham: Springer International Publishing.

Wasserman, S., & Faust, K. (1994). *Social Network Analysis: Methods and Applications*. Cambridge: Cambridge University Press.

Watershed Management Group & Pima County Regional Flood Control District. (2015). Solving Flooding Challenges with Green Stormwater Infrastructure in the Airport Wash Area.

Watershed Management Group. (n.d.). *History of Watershed Management Group 2014*.

Weible, C. M. (2007). An Advocacy Coalition Framework Approach to Stakeholder Analysis: Understanding the Political Context of California Marine Protected Area Policy. *Journal of Public Administration Research and Theory, 17*(1), 95–117.

Weible, C. M., Sabatier, P. A., & McQueen, K. (2009). Themes and Variations: Taking Stock of the Advocacy Coalition Framework. *Policy Studies Journal: The Journal of the Policy Studies Organization, 37*(1), 121–140.

WRRC. (2008). *NEMO Watershed-Based Plan Santa Cruz Watershed*.

Yi, H., & Scholz, J. T. (2016). Policy Networks in Complex Governance Subsystems: Observing and Comparing Hyperlink, Media, and Partnership Networks. *Policy Studies Journal: The Journal of the Policy Studies Organization, 44*(3), 248–279.

12

Conclusions

Karin Ingold and Manuel Fischer

This book includes nine case studies and social network applications pertaining to water issues and challenges in different parts of the world. The nine case study chapters cover issues related, but not restricted, to international conflicts, water quality and quantity challenges, urban water management, biodiversity and ecosystem conservation. This book demonstrates that water governance—and, we would argue, complex governance situations—can fruitfully be conceived of as a network of a variety of actors or issues, or both together. In this conclusion, we briefly outline the main findings of the case study chapters, as well as general contributions the chapters as a whole make to academic theory and methodological approaches to dealing with complex water governance. We also offer some input for practitioners.

K. Ingold (✉) • M. Fischer
University of Bern, Bern, Switzerland

Federal Institute of Aquatic Science and Technology, Eawag,
Dübendorf, Switzerland
e-mail: karin.ingold@ipw.unibe.ch; manuel.fischer@eawag.ch

© The Author(s) 2020   **321**
M. Fischer, K. Ingold (eds.), *Networks in Water Governance*, Palgrave Studies in Water
Governance: Policy and Practice, https://doi.org/10.1007/978-3-030-46769-2_12

We identify two main challenges to water governance (see Fischer and Ingold, Chap. 2 in this book): the first one is related to the fact that water issues are very often multi-level, cross-sectoral, trans-boundary, and that causes and effects of water related problems are disentangled in space and time. For water governance, this means that actors need to coordinate actions across sectors, administrative boundaries, and multiple levels. The chapters by Robbins and Lubell, Mancilla and Bodin, Herzog and Ingold, or Ebrahimiazarkharan et al. (all this book) explicitly confirm the challenge of designing coordinated action to solve water issues across borders, different geographical territories, up- and downstream areas of a water catchment. Further below, we summarize some factors that might overcome multi-level and trans-boundary barriers to coordination in water governance.

Angst and Fischer as well as Bell and Henry (both this book), still related to this first challenge, address the issue of cross-sectoral coordination and focus on some specific actors such as coordinators, brokers, or entrepreneurs in water policy networks that are able to overcome sectoral silos.

The second challenge we identify in this book is related to the first one: addressing the first challenge means, including a variety of actors into water governance arrangements. But this in turn means that diverse types of actors come together, very often with different stakes, interests, and resources. In consequence, water governance arrangements can rather be characterized as potentially conflictive and heterogeneous, where collaboration and coordination is a challenge. This fact is shown by the analysis of Hollway (this book), who studies international conflict and coordination. But also Mancilla and Bodin, as well as Koebele et al. (both this book) dedicate their studies to the issue of beliefs and interests: they identify so-called advocacy coalitions that consist of members sharing similar beliefs and worldviews about how to solve the water-related challenges, but who are in opposition to other actors and their beliefs.

In sum, this book shows, on the one side, that these water-related challenges we outlined at the beginning (Chap. 2) are very often present and observable. But the chapters also make a contribution as to how to overcome these challenges. Finally, we confirm that applying social network analysis for the study of water governance can be useful to exactly identify where the conflict lines are, and also to highlight the potential for collaboration and coordination across sectors, boundaries, levels, and scales.

The book is structured into a first part that deals with network fragmentation, a second set of three chapters that show how fragmentation can be overcome, and a third part focusing on actors that play specific roles in network governance structures. Overall, the empirical case study chapters then generally address two main research questions related to these challenges and the respective book structure: First, authors ask which actors are particularly important and occupy central positions within networks, and how such central positions are distributed within networks or parts thereof. Second, authors ask why actors in water governance networks interact with some actors, but not with others. In what follows, we briefly summarize the key findings of each chapter, present some overall conclusions, and finally lay out insights for practice.

Key Findings per Chapter

The chapter by Robbins and Lubell analyzes drivers for network segregation, as well as reasons for tie creation. They illustrate these mechanisms through the case of the Spiny Lobster Initiative (SLI) in Honduras. The authors find that homophily, in terms of territoriality and sectoral affiliations, matters for tie creation. This means that cross-sectoral and trans-boundary coordination among actors is still a challenge. Many of today's water challenges are trans-boundary or cross-sectoral, and overcoming geographical or sectoral barriers is a particular challenge. There is a broad body of literature dealing with the role of forums such as the SLI to overcome such barriers (Berardo and Scholz 2010; Fischer and Maag 2019; Herzog and Ingold 2019). However, in the case studied here, the SLI initiative was only partially able to overcome network segregations.

In his chapter, Hollway studies international water-related cooperation and conflict. The chapter represents the first demonstration of coevolving, signed DyNAMs, as well as one of the first empirical and theoretical emphases by an actor-oriented network model of the rate part of the model. The author finds that network configurations matter also when controlling for other, non-network variables based on the general literature on conflict and cooperation. The key finding is that past coop-eration accelerates the establishment of cooperative ties, whereas conflict slows this process down. International water governance is thus heavily shaped by network events from the past.

Angst and Fischer outline a two-mode network between actors and issues. They first identify five subsystems dealing with different aspects of the larger topic of water governance. The chapter then focuses on so-called connectors with a potentially strong influence on network struc-ture, dynamics, and outputs. Within-subsystem connectors are central in one subsystem and have the capacity to shape decisions therein. Even more interestingly, for overcoming the segregation as outlined by Robbins and Lubell (this book), between-subsystem connectors are working on a variety of issues from diverse subsystems. Being able to play such a role, however, strongly depends on actors' resources.

In their chapter, Mancilla and Bodin analyze a water governance case from a Brazilian watershed. They analyze coalition formation and power relations among public and private actors, and show that actors have the tendency to interact with powerful and influential others, and that power oftentimes outweighs the importance of beliefs and ideologies. The authors conclude that similar ideologies do not always account for tie creation, and that actors may even have the tendency to switch advocacy coalitions. Similar to other studies (e.g. Calanni et al. 2014), the authors show that for regional water governance, power can be a more stable tie predictor than ideology.

Koebele et al. analyze the Lake Tahoe Basin environmental governance subsystem for the period 2004 until 2015. The authors confirm their key assumption that the Lake Tahoe water governance subsystem is multifac-eted, and that no single network type can capture the different faces of its complexity. They draw three different networks of beliefs, policy posi-tions, and interactions from the newspaper data and observe that actors can take very different positions in each of the networks. Still, advocacy coalitions of actors sharing the same beliefs have the tendency to coordi-nate their actions more than actors that are not in the same coalition. Thus, ideology can (Koebele et al., this book) but does not have to be (Mancilla and Bodin, this book) a predictor of joint coordination in water governance.

In their chapter, Herzog and Ingold study a relatively new water quality issue in three Rhine catchment areas: micro-pollutants stemming from households, agriculture, or the (chemical) industry. They combine policy studies with the social-ecological systems (SES) framework and network analysis in order to see why and when public and private actors coordinate to tackle this issue. Whereas Basel and the Ruhr sub-catchments have measures of regulation in place, in the Moselle region, the issue just entered the political agenda. The core assumption is that the three coordination networks look very different and account for the difference in policy outputs and outcomes. The analysis, however, shows that the networks are similar, and that the (technical) issue of micro-pollutants, as well as the problem pressure, might lead to coordination taking place across different sectors and levels.

Ebrahimiazarkharan et al. dedicate their chapter to local stakeholders and their trust and cooperation relations with each other. They focus on three regions in the Taleghan watershed in Iran and its upstream, center, and downstream catchments. The authors state that the identification of key actors and the enhanced involvement of local stakeholders in the process could improve this situation of inappropriate management and water-related conflicts. Their results show that, in contrast to the upstream and central region, the downstream region suffered from low network density. Some key connectors between the different actors could be identified: authorities with high degree and betweenness centralities have the potential to increase the quality of water management and to overcome issues lacking social capital and sustainability.

The chapter by Olivier et al. studies the largest water provider is operating under a Filtration Avoidance Determination granted by the Environmental Protection Agency in the US: New York City delivers 1.4 billion gallons of largely unfiltered water daily to 9 million people. The authors create networks between actors and rules by analyzing the 3000 plus rules that constitute the New York City Watersheds regional governing arrangement. They draw on this corpus of rules to identify how rules define behavioral expectations in terms of relations between the parties to the agreement. They find that rules establishing credible commitments create network structures that differ from the ones created through rules guiding the provision of public goods, indicating that actors design institutions anticipating that different kinds of collective action problems require different governance structures.

In the last empirical chapter, Bell and Henry rely on text coding to draw a two-mode network between actors and water-related innovations in Tucson, Arizona, and analyze it as a one-mode network, with actors related through joint event participation. In their next step, Bell and Henry focus on the most central actors: namely, ones who are able to connect to a multitude of actors at different events (degree centrality), or to connect otherwise disconnected others (betweenness centrality). This comes close to the connector concept used in Angst and Fischer (this book). Additionally, Bell and Henry code media accounts, where actors were explicitly mentioned as being brokers or entrepreneurs, and compare this data to network measures of degree and betweenness centrality. Results show that both degree and betweenness correlate with the explicit report of importance in the media, but, for brokers and entrepreneurs, betweenness seems to grasp better its connector role than degree centrality.

Learnings from and for the Application of Network Concepts and Measures to Water Governance

The first research question we outline in Chap. 2 asks about the role of individual key actors in water governance. Most of those key actors in networks are identified through degree, betweenness, or other centrality measures (see Bell and Henry, Angst and Fischer, Ebrahimiazarkharan et al.; this book). This means, those actors connect to many others, or connect different parts of the network that would otherwise be disconnected (see Bell and Henry, this book). They thus occupy an advantageous position either to advocate compromise between conflicting parties, or to push solutions forward. In a water governance situation, this means that those actors might be able to overcome the different sectoral divides and propose solutions grounded on principles, such as sustainability or integrated water resources management (see Ebrahimiazarkharan et al., this book). Studies in different chapters identified such connectors, brokers, and entrepreneurs—all examples of single key actors holding the network together, or being able to lead and mobilize actors for new solutions. However, some questions remain: For example, Ebrahimiazarkharan et al. (this book) identify the most central actors in the three investigated areas of the Taleghan watershed, and Herzog and Ingold (this book) perform an analogous task for the three sub-catchments of the Rhine. Yet, the mere identification of central actors does not yet say much about their role and impact on governance outcomes. This is why Bell and Henry (this book), for instance, opt for validation and compare network data to media analysis about brokers and entrepreneurs. In conclusion, network measures, such as different centralities, help identify most active or connecting actors in a governance setting, but case knowledge or qualitative data are needed to know the concrete impact of those actors on procedural or substantive outcomes in water governance. Ebrahimiazarkharan et al. (this book) use a slightly other approach and mention that knowing or identifying the most central actors creates an opportunity for policy recommendations and outreach activities: showing network results to the actors and stakeholders involved in this network, or to (local or national) authorities, might be an effective tool in shaping future governance decisions (also see next section).

One particular type of actor that is able to positively impact coordination in water governance networks are authorities and state actors. Along with Börzel and Risse (2010), network effectiveness is enhanced in "the shadow of hierarchy" displayed by state actors. For example, in the Spiny Lobster network, analyzed by Robbins and Lubell (this book), authorities and government actors are the targets of tie creation. This network is hierarchical and centralized, with a small number of actors at the core of interaction. Following the authors, this fact might actually facilitate effective coordination among actors and the efficient spread of information. Also, Herzog and Ingold (this book) find a high number of actors among the most central ones in all three coordination networks along the Rhine River. The rather technical issue of micro-pollutant regulation in surface waters is thus led and coordinated by state actors, who seem to have done a good job in involving further public and private actors: in the three cases, networks are rather dense without much fragmentation. In sum, and as one answer to the first research question, authorities still play an important role in water governance and can facilitate the quality of structural and procedural aspects of network governance.

In line with the second research question (why and how do actors coordinate) and the first challenge (fragmentation) as outlined above, the remaining puzzle is: how can ideological, geographical, or organizational divides and cleavages be overcome in order to reduce segregation and foster coordination in water governance across levels, borders, and sectors? While the first three empirical case studies tackle this issue explicitly, most chapters, in one way or another, deal with fragmentation or segregation and the question of how to reduce it. Herzog and Ingold (this book) show, through a faction analysis, that the fragmentation in the water quality coordination network in three different sub-systems along the Rhine had different reasons. While in the French/Luxembourg case of the Moselle, actors form a divide along territoriality, in the cross-border region of Basel, Switzerland, and in the Ruhr-catchment in Germany, actor type seems most decisive for collaboration. This is in line with Robbins and Lubell (this book), who also conclude that either geographical or actor type homophily make actors coordinate action; and that overcoming sectoral logics and territorial borders seems still one of the biggest challenges in water governance. Also in the Taleghan watershed in Iran, analyzed by Ebrahimiazarkharan et al. (this book), upstream and downstream divides were hard to overcome and cross-sectoral and multi-level water governance were still a challenge to establish. Koebele et al. (this book) showed some ideological segregation in their analysis of the Lake Tahoe basin: interestingly, also within-coalition conflicts were apparent. The pro-environment coalition, for instance, did not agree upon the effectiveness of the concrete policy program, even though they are all in favor of more ecological performance.

Another contribution of this book is thus to show different pathways on how fragmentation can be overcome and coordination fostered.

Very generally speaking, fragmentation and segregation can be overcome through networks, but there is not one type of network ties that matters above all others. Rather, there are many different types of ties in networks, and they combine and influence each other in different ways. Ebrahimiazarkharan et al. (this book) come to the conclusion that, in the three investigated sub-catchments of the Taleghan watershed in Iran, trust and cooperation relations are established differently and one does not automatically induce the other (results not displayed in the book). For instance, in most of the cases, trust relations show a higher density and reciprocity than cooperation. Even though the two might correlate, trust does not immediately lead to cooperation on water-related problems nor vice versa. Thus factors other than interpersonal relations also play an important role in how to set up coordination within a watershed, such as institutions, or path dependencies, or actors' affiliations. For instance, Robbins and Lubell (this book) show that organizations belonging to the same type or same geographical area have the tendency to coordinate actions. Territorial and actor type homophily can shape or facilitate coordination. Similarly, also Mancilla and Bodin (this book) conclude this jurisdictional homophily effect: except for private actors, organizations tend to coordinate actions with territorial peers.

Coming back to the type of tie that matters: Hollway (this book) shows the interrelation between cooperation and conflict, but also impacts of past ties on the current establishment of relationships. International water-related cooperation between states is thereby shaped by both cooperation and conflict from the past. The first acts as accelerator, whereas the latter slows the network development of cooperative ties down.

In sum, and as a first answer to the second research question, different chapters showed that actor type or territorial homophily, trust and past relations might positively impact coordination in water governance.

A final driver for coordination among actors is institutions. Although none of the case study chapters, nor our general approach, has a strong and explicit focus on institutions, they are present in at least two ways. First, institutions play a role in governance settings through rules. Olivier et al. (this book) showed that different types of rules predict different forms of interactions by actors involved in water-related issues. They assume, for instance, that protocols related to power-sharing agreements (credible commitments) or providing for complex (in contrast to simple public) goods are likely to outline multiple forms and contexts of interactions among the actors involved, in order to hedge against nonperformance. They could only partially confirm this assumption, but their analysis definitely shows that different problems and types of goods outline diverse forms of (expected) interactions among actors. If actors follow the rules, and if rules are designed with the goal of enhancing cross-sectoral and multi-level coordination, network cohesion in water governance might increase.

A second form of institutions that are important for understanding governance networks are forums, defined as platforms where public and private actors, experts, and scientists (regularly) gather together can enhance coordination. Depending on the nature, and the setup, of the forum, they have the potential to include actors that otherwise would rarely meet. In this regard, forums can be important learning places, as well as also agenda-setters (Fischer and Leifeld 2015). Herzog and Ingold (this book) did not explicitly study the role of the International Commission for the Protection of the Rhine (ICPR) that includes all member states of the catchment area, as well as further actors from the private sector, drinking water providers, waste water treatment associations, experts, and scientific institutes. But the high degree of coordination in all three sub-catchments (Ruhr, Basel, and Moselle), all belonging to different countries, might have been strongly impacted by the fact that the regulation of micro-pollutants was already "prepared" by the ICPR, and that actors know each other from different meetings and workshops organized therein. Of course, not all forums necessarily have such an impact. Again, in the Spiny Lobster Initiative, analyzed by Robbins and Lubell (this book), the respective forum was not particularly decisive for tie creation. A nuanced view on forums is needed; what our book shows is that forums designed in a trans-national, cross-sectoral, and multi-level manner may have the potential to facilitate governance in water networks.

Recommendations for Academia and Practice

In terms of recommendation, our first message is directed at researchers interested in applying SNA to water-related issues. Water governance cases might be more complicated than other fields when it comes to appropriately designing network boundaries, given the important cross-sectoral, trans-boundary, and multi-level aspects (see also last part of

Fischer and Ingold, Chap. 2, this book). Network boundaries are most often drawn around the actors that are involved in water management and decision-making. As the applications in this book have shown, actors potentially involved in water management or policy issues can be diverse: besides authorities and state actors, this can include private firms, public private partnerships, non-governmental organizations, consumer organizations and trade unions, science and other experts, and many more. Not including relevant actors would likely create a distorted impression about the water governance network under investigation. We thus propose that it is thus less of a problem to—*a priori*—include organizations that prove to be extraneous than to fail to include an important one. Including "unnecessary" organizations would be a self-correcting issue, because as soon as an organization is deemed to be irrelevant in the governance arrangement, network metrics would declare it as an isolate (no connection at all), or an actor at the periphery of the network (only related to one or two actors in the center). Another way to be sure that the "right" actors are considered for the analysis is to combine several approaches together: different chapters have outlined the consultation of protocols, expert interviews, or asking the already identified actors about further organizations to include (e.g. snowball sampling).

Also for practitioners, for decision-makers, and for experts involved in water management, the identification of key actors is crucial. As already mentioned in the section above, network analysis does not have to be limited to being an academic exercise—it can provide a meaningful tool for identifying important brokers or connectors. Results from social network analysis, and especially network visualizations, can be presented to decision-makers and practitioners who are interested to know how to enhance coordination, or whom to contact to reach the maximum of stakeholders in their region. Actors with high degree centrality can act as important multipliers, and authorities might want to use them to diffuse important information to many others. Alternatively, organizations with high betweenness centrality connect otherwise disconnected parts of the governance network. If an agency is interested in reaching all parts of the management arrangement, contacting those with high betweenness centrality can be an effective option.

Robbins and Lubell (this book), as well as many other case study chapters, show here that connecting actors from other territories and other sectors is still a major challenge in the applied water governance situations. Mancilla and Bodin (this book) show that mainly firms, in contrast to public actors, are able and willing to "go abroad" and connect to others from other jurisdictions. This is in line with the study by Fischer and Maag (2019) that comes to the conclusion that mainly firms are interested in participating in cross-sectoral, public-private forums to exchange experiences and to impact the political agenda. In conclusion, it seems relevant and wise for authorities to include firms and private actors in water governance decisions, as they have the potential and interest to push innovations, and to think across borders and beyond uni-dimensional processes.

To push new ideas and create policy innovations, not only the involvement of the private sector but also the general understanding and acceptance of the problem seem crucial. Koebele et al. (this book) mentioned that all actors, also members belonging to opposing coalitions (pro-economy versus pro-environment), agreed upon the fact that the Lake Tahoe region needs to be ecologically conserved. A joint understanding of the problem and a general understanding of what could be done in terms of policy options helps to find appropriate solutions. This book shows the potential of network analysis to be applied to a variety of water governance challenges and thereby help to identify communication channels in order to enhance information exchange and trust among actors. This in turn does not only serve the purpose in gaining insights in how governance arrangements work but can be used as a first step in order to identify communication gaps, learning potential, and the formulation of joint commitments to solve water-related problems.

References

Berardo, R., & Scholz, J. T. (2010). Self-Organizing Policy Networks: Risk, Partner Selection, and Cooperation in Estuaries. *American Journal of Political Science, 54*(3), 632–649.

Börzel, T., & Risse, T. (2010). Governance Without a State: Can it Work? *Regulation & Governance, 4*, 113–134.

Calanni, J. C., Siddiki, S. N., Weible, C. M., & Leach, W. D. (2014). Explaining Coordination in Collaborative Partnerships and Clarifying the Scope of the Belief Homophily Hypothesis. *Journal of Public Administration Research and Theory, 25*(3), 901–927.

Fischer, M., & Leifeld, P. (2015). Policy Forums: Why Do They Exist and What Are They Used For? *Policy Sciences, 48*(3), 363–382.

Fischer, M., & Maag, S. (2019). Why Are Cross-Sectoral Forums Important to Actors? Forum Contributions to Cooperation, Learning, and Resource Distribution. *Policy Studies Journal, Online.* https://doi.org/10.1111/psj.12310.

Herzog, L. M., & Ingold, K. (2019). Threat to Common-Pool Resources and the Importance of Forums: On the Emergence of Cooperation in CPR Problem Settings. *Policy Studies Journal, Online.* https://doi.org/10.1111/psj.12308.

CPSIA information can be obtained
at www.ICGtesting.com
Printed in the USA
LVHW081941151120
671754LV00001B/51

.